NATURE ALWAYS CREATES THE BEST OF ALL OPTIONS.

ARISTOTLE

ONE SCIENTIFIC EPOCH ENDED AND ANOTHER BEGAN WITH JAMES CLERK MAXWELL.

ALBERT EINSTEIN

THE TRUTH ALWAYS TURNS OUT TO BE SIMPLER THAN YOU THOUGHT.

RICHARD FEYNMAN

JAKOB SCHWICHTENBERG

NO-NONSENSE ELECTRODYNAMICS

NO-NONSENSE BOOKS

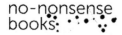

First printing, September 2020

Copyright © 2020 Jakob Schwichtenberg
All rights reserved. No part of this publication may be reproduced, stored in, or introduced into a retrieval system, or transmitted in any form or by any means (electronic, mechanical, photocopying, recording, or otherwise) without prior written permission.

UNIQUE ID: 452E0192CDCFF67CB9AEA345AF09225B67F5553C103453C891C8EC0569F67E9E
Each copy of No-Nonsense Electrodynamics has a unique ID which helps to prevent illegal sharing.

BOOK EDITION: 1.6

Dedicated to my parents

Preface

There are already dozens of textbooks on electrodynamics. So why another one?

First of all, almost all existing textbooks follow the same "well-established route" which is inspired by the historical developments. The main idea is to start by discussing lots of experimental facts and only afterwards put these puzzle pieces together. This way the reader is slowly guided towards Maxwell's general theory of electrodynamics.[1]

[1] In this sense, this is a bottom-up approach.

However, I don't think this approach is the most effective and certainly not the most entertaining one. Without seeing the bigger picture, discussing individual electric and magnetic phenomena can be quite confusing and makes electrodynamics appear more complicated than it really is. Moreover, readers are often left in the dark regarding which statements hold in general and which only in special situations. In addition, books which follow such a bottom-up approach usually spend most of the time at the "bottom" and only near the very end get to the "top". This often means that the general theory is not discussed in detail and many questions remain unanswered.

In contrast, this book follows a top-down approach. We start by discussing the general theory and only afterwards discuss various special cases. All fundamental aspects of electrodynamics are described in the first chapter and put into context. This allows the reader to see the bigger picture right away and

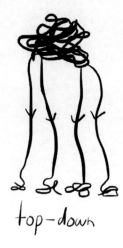

top–down

immediately makes clear which statements hold in general. In addition, there is no risk that students give up before they get to the most important stuff.

Secondly, most other electrodynamics textbooks try to do a lot at once. For example, it's not uncommon that in addition to Maxwell's general theory, dozens of applications, edge cases, advanced topics, historical developments or even biographies of the most important contributors are discussed. I think this is problematic because, as the saying goes, if you try to be good at everything, you will not be great at anything.

In contrast, this book focuses solely on the fundamental aspects of electrodynamics.[2] This narrow focus allows us to discuss all the important concepts several times from various perspectives. A clear advantage of this approach is that the reader has multiple chances to understand a given concept while in a normal textbook, the reader immediately has a problem when a passage is not understood perfectly.[3] A second advantage of our narrow focus is that it minimizes the risk of unnecessarily confusing the reader. Like all other fundamental theories, electrodynamics is, at its heart, quite simple. However, using it to describe complicated systems is far from easy and this is where most of the difficulties usually arise.[4] Therefore, restricting ourselves to the fundamentals allows us to introduce electrodynamics as gently as possible.[5]

While this alone may already justify the publication of another electrodynamics textbook, there are a few other things which make this book different:

▷ First of all, it wasn't written by a professor. So this book is by no means an authoritative reference. It's more like a casual conversation with a more experienced student who shares with you everything he wished he had known earlier. I'm convinced that someone who has just recently learned the topic can explain it much better than someone who learned it decades ago. Many textbooks are hard to understand, not because the subject is difficult, but because the author can't remember what it's like to be a beginner.

---

[2] Applications are only discussed insofar as they help to deepen our understanding of the fundamental concepts and not as an end in themselves. In addition, there are already dozens of great books which discuss applications or other special topics in great detail. Some of the best ones are recommended in Chapter 9.

[3] In a normal textbook each topic is only introduced once. As a result, later chapters become harder and harder to understand without a full understanding of all previous chapters. Moreover, it's easy to become discouraged when a few passages are not perfectly clear since you know that you need the knowledge to understand later chapters.

[4] Most of the difficulties are really mathematics problems not physics problems anyway. For example, solving a difficult integral or solving a given differential equation.

[5] While advanced applications are, of course, important, they are not essential to understand the fundamentals of electrodynamics. As already mentioned above, there are great books which focus on specific applications. After you've developed a solid understanding of the fundamentals, it's far easier to learn more about those applications you're interested in.

▷ Another aspect that makes this book unique is that it contains lots of idiosyncratic hand-drawn illustrations. Usually, textbooks include very few pictures since drawing them is either a lot of work or expensive. However, drawing figures is only a lot of work if you are a perfectionist. The images in this book are not as pretty as the pictures in a typical textbook since I firmly believe that *lots* of imperfect illustrations are much better than a few perfect ones. The goal of this book, after all, is that you'll understand electrodynamics and not that I win prizes for my pretty illustrations.

▷ Moreover, my only goal with this book was to write the most student-friendly electrodynamics textbook and not, for example, to build my reputation. Too many books are unnecessarily complicated because if a book is hard to understand it makes the author appear smarter.[6] Nothing is assumed to be "obvious" or "easy to see". Moreover, calculations are done step-by-step and annotated to help you understand faster.

[6] To quote C. Lanczos: "Many of the scientific treatises of today are formulated in a half-mystical language, as though to impress the reader with the uncomfortable feeling that he is in the permanent presence of a superman."

So, without any further ado, let's begin. I hope you enjoy reading this book as much as I have enjoyed writing it.

Karlsruhe, October 2018 *Jakob Schwichtenberg*

PS: I update the book regularly based on reader feedback. So if you find an error, I would appreciate a short email to errors@jakobschwichtenberg.com.

Acknowledgments: Special thanks to Michael Havrilla, Tom Harper, Luke Durant, Steve Burk, Alex Huang, Jonathan Hobson, Vicente Aboites, Xavier Constant, Fabian Waetermans, and Phil Connolly for reporting several typos. Moreover, many thanks are due to Jacob Ayres for his careful proofreading and countless invaluable suggestions.

Before we dive in, we need to talk about two things. The first one is the following crucial question:

Why should you care about Electrodynamics?

First of all, electrodynamics correctly describes the behavior of one of only four known fundamental interactions.[7]

[7] This is true at least on a macroscopic level. Electromagnetic interactions of elementary particles are described by quantum electrodynamics.

▷ At large (cosmological) scales, gravity is the most important of the four interactions.

▷ At tiny scales, strong and weak interactions are responsible for most of the interesting phenomena. For example, protons are held together by the strong force and atoms decay through weak interactions.

▷ However, in between these two extremes it's electromagnetic interactions which hold sway.

In particular, electrodynamics not only allows us to understand how electricity and magnetism come about, but also what light is and how it travels.

In addition, a solid understanding of electrodynamics is essential to understand the theories describing all other fundamental interactions. Moreover, the equations describing electromagnetic interactions of macroscopic objects are exactly the same as the equations governing electromagnetic interactions of elementary particles.[8] This means that the knowledge gathered by studying electromagnetic interactions on a macroscopic level helps us immediately to understand what happens on a more fundamental level.

[8] The equations describing the other fundamental interactions are extremely similar and can be understood much easier once the equations of electrodynamics are understood.

Moreover, electrodynamics is an ideal playground to understand several of the most important concepts underlying modern physics in a simplified setup. In particular, in the context of electrodynamics we can understand:

▷ What gauge symmetry is.
▷ What a gauge field is. How we can interpret it geometrically and why gauge fields are responsible for fundamental interactions.
▷ What special relativity is all about.

Lastly, electrodynamics is prototypical for what progress in physics means. Before it was developed, even a genius like Leonhard Euler stated:

"The subject I am going to recommend to your attention almost terrifies me. The variety it presents is immense, and the enumeration of facts serves to confound rather than to inform. The subject I mean is electricity."

Nowadays, all we have to do to understand the multitude of electric and magnetic phenomena is to study five short equations and you don't have to be a genius to master electricity.

Formulated differently, electrodynamics is certainly one of the most important scientific works of all time. This is especially true if we use David Hilbert's criterion:

"One can measure the importance of a scientific work by the number of earlier publications rendered superfluous by it."

Before Maxwell developed electrodynamics in its modern form, there were hundreds of publications, each describing a different electric and magnetic phenomena or, for example, properties of light. Nowadays, we only need Maxwell's equations to describe all of this.

The second thing that we need to talk about is the meaning of a few special symbols which we will use in the following chapters.

Notation

▷ Three dots in front of an equation ∴ mean "therefore", i.e., that this line follows directly from the previous one:

$$\omega = \frac{E}{\hbar}$$
$$\therefore \quad E = \hbar\omega \, .$$

This helps to make it clear that we are *not* dealing with a system of equations.

▷ Three horizontal lines $\equiv$ indicate that we are dealing with a definition.

▷ The symbol $\stackrel{!}{=}$ means "has to be", i.e. indicates that we are dealing with a condition.

▷ The most important equations, statements and results are highlighted like this:

$$\boxed{\nabla^2 \vec{E} = \mu_0 \epsilon_0 \frac{\partial^2}{\partial t^2} \vec{E}}$$

▷ We often use the shorthand notation $\partial_i \equiv \frac{\partial}{\partial i}$ for the partial derivative where $i \in \{x, y, z\}$.

▷ We write $\vec{x}$ as the argument of a function $f(\vec{x})$ as a shorthand notation for $f(\vec{x}) = f(x, y, z)$. Moreover, often the explicit argument is suppressed altogether to unclutter the notation.

▷ The symbol $\oint$ is used for integrals over *closed* surfaces and *closed* paths.

▷ We denote the components of a vector $\vec{v}$ by v_i.

▷ Whenever the word "charge" is mentioned without further context, we always mean electric charge.[9]

▷ If an index occurs twice, a sum is implicitly assumed:

$$\sum_{i=1}^{3} a_i b_i = a_i b_i = a_1 b_1 + a_2 b_2 + a_3 b_3 ,$$

[9] In a more general context, it could also mean weak charge, color charge etc.

but
$$\sum_{i=1}^{3} a_i b_j = a_1 b_j + a_2 b_j + a_3 b_j \neq a_i b_j.$$
This is known as Einstein's summation convention.

▷ Greek indices like μ, ν or σ, are always summed from 0 to 3:
$$x_\mu y_\mu \equiv \sum_{\mu=0}^{3} x_\mu y_\mu.$$

▷ In contrast, Roman indices like i, j, k are always summed from 1 to 3:
$$x_i x_i \equiv \sum_{i=1}^{3} x_i x_i.$$

▷ Basis vectors are always denoted by $\vec{e}_i$. For example, we use $\vec{e}_x, \vec{e}_y, \vec{e}_z$ for the basis vector in a Cartesian coordinate system and $\vec{e}_r, \vec{e}_\varphi, \vec{e}_\theta$ for the basis vectors in a spherical coordinate system.

▷ δ_{ij} denotes the Kronecker delta, which is defined as follows:
$$\delta_{ij} = \begin{cases} 1 & \text{if } i = j \\ 0 & \text{if } i \neq j \end{cases}$$

▷ ϵ_{ijk} denotes the three dimensional Levi-Civita symbol:
$$\epsilon_{ijk} = \begin{cases} 1 & \text{if } (i,j,k) = \{(1,2,3),(2,3,1),(3,1,2)\} \\ 0 & \text{if } i = j \text{ or } j = k \text{ or } k = i \\ -1 & \text{if } (i,j,k) = \{(1,3,2),(3,2,1),(2,1,3)\} \end{cases}$$

―――――――

That's it. We are ready to dive in. (After a short look at the table of contents).

Contents

1 Birds-Eye View of Electrodynamics   19

Part I What Everybody Ought to Know About Electrodynamics

2 **Fundamental Concepts**   31
    2.1 Electric charge 32
    2.2 Charge density 35
    2.3 The electric current 37
    2.4 The electromagnetic field 42
    2.5 The electromagnetic potential 48
    2.6 Summary 50

3 **Fundamental Equations**   53
    3.1 The continuity equation 54
    3.2 The Lorentz force law 57
    3.3 Gauss's law for the electric field 61
    3.4 Gauss's law for the magnetic field 69
    3.5 Faraday's law 74
    3.6 The Ampere-Maxwell law 79
    3.7 The wave equations 82
    3.8 Summary 84

Part II Essential Electrodynamical Systems and Tools

4 **Electrostatics and Magnetostatics**   95
    4.1 Electrostatic Systems 101

		4.1.1	Electric field of a single point charge	101
		4.1.2	Electric field of a sphere	106
		4.1.3	Electric field of an electric dipole	110
		4.1.4	Electric field of more complicated charge distributions	111
		4.1.5	Charged object in a static electric field	114
		4.1.6	Further Systems	117
	4.2	Magnetostatic Systems		123
		4.2.1	Magnetic field of a wire	123
		4.2.2	Charged object in a static magnetic field	125
		4.2.3	Further Systems	130

5 Electrodynamics   **133**

    5.1 An explicit solution of the wave equation 134
    5.2 Corresponding solution of the magnetic wave equation 136
    5.3 General solutions of the wave equations 138
    5.4 Basic properties of electromagnetic waves 143
    5.5 Advanced properties of electromagnetic waves . . . 149

Part III Get an Understanding of Electrodynamics You Can Be Proud Of

6 Special Relativity   **163**

    6.1 The origin of Maxwell's equations 169

7 Gauge Symmetry   **175**

8 Electrodynamics as a Gauge Theory   **181**

    8.1 Symmetries intuitively 182
        8.1.1 Global vs. local symmetries 183
    8.2 A toy gauge theory 184
        8.2.1 Active vs. passive transformations 186
        8.2.2 Symmetries vs. redundancies 188
    8.3 Gauge dynamics 190
        8.3.1 Mathematical description of the toy model . 194
        8.3.2 Gauge rules 200
    8.4 Gauge symmetry in physics 203
        8.4.1 Gauge symmetry in Quantum Mechanics . . 203
        8.4.2 Gauge Symmetry in Electrodynamics 206

	8.4.3 Putting the puzzle pieces together 208
8.5	Gauge symmetries mathematically 210
	8.5.1 Gauge connections in Quantum Mechanics and the toy model 216

9 Further Reading Recommendations   219

Part IV Appendices

A Vector Calculus   227
- A.1 Scalars, Vectors, Tensors 233
- A.2 The dot product 235
- A.3 The cross product 238
- A.4 Fields 240
- A.5 Line integral 244
- A.6 Path integral 247
  - A.6.1 Tangent vector 250
- A.7 Circulation integral 253
- A.8 Surface integral 254
  - A.8.1 Example: surface integral 256
- A.9 Flux Integral 257
- A.10 Gradient 262
- A.11 Divergence 264
- A.12 Curl 270
- A.13 The fundamental theorem for gradients 276
- A.14 The fundamental theorem for divergences a.k.a. Gauss's theorem 278
- A.15 The fundamental theorem for curls a.k.a. Stokes' theorem 281
- A.16 Vector identities 284
  - A.16.1 The Bianchi identity 287
  - A.16.2 Summary of vector identities 289
- A.17 Index notation and Maxwell's equations 290
  - A.17.1 Electrodynamical Lagrangian 294

B Taylor Expansion   297

C Delta Distribution   301

Bibliography   305

1

Birds-Eye View of Electrodynamics

As mentioned in the preface, electrodynamics is at its heart, like most other theories in physics, quite simple. However, certain applications can be extremely complicated. For this reason it's easy to lose the forest for the trees. To prevent this, we start this book by talking about all fundamental notions and concepts and putting them into context. Afterwards, we will talk about the various concepts in more detail and gradually refine our understanding until we are ready for real-world applications.

There are three kinds of puzzle pieces that we need to master in order to get a proper understanding of electrodynamics:

1. Concepts (electric charge, electric current, charge density, current density, electric and magnetic field, electromagnetic potential, electromagnetic waves)
2. Equations (continuity equation, Maxwell's equations, Lorentz force law, wave equations)

3. Tools (vector calculus, the delta distribution)

In addition, there are lots of formulas which follow from these fundamental equations[1] and several important advanced mathematical tools[2] and concepts[3]. These are important for specific applications and will be discussed when they are needed, not in advance.

Don't worry if not everything is immediately clear in this chapter. Our goal is solely to get an overview and each idea discussed in this chapter will be discussed later in more detail.

[1] For example, Coulomb's Law and the Biot-Savart Law. In addition, take note that only Maxwell's equations and the Lorentz force law are really fundamentals. The wave equations and the continuity equation can be derived using them.

[2] Greens functions, method of image charges, multipole expansion.

[3] Polarization, Poynting vector, gauge invariance

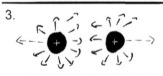

First of all, what is our goal in electrodynamics?

The short version is:

> We want to understand the interplay between electrically charged objects and the electromagnetic field.

This rather broad goal consists of four smaller goals:

1. We want to describe how charged objects behave in the presence of the electromagnetic field.
2. We want to describe how charged objects influence the electromagnetic field.
3. We want to describe how charged objects interact with each other.
4. We want to understand how the electromagnetic field behaves when there are no charged objects present.[4]

All this is accomplished by **Maxwell's equations**[5]

[4] Typically, in this context we speak about electromagnetic radiation. The most famous example of electromagnetic radiation is light.

[5] Be assured that these equations are not really as complicated as they may seem at first glance. In the next chapter we will see that Maxwell's equations are quite natural statements translated into a mathematical form.

$$\partial^\mu F_{\mu\nu} = \mu_0 J_\nu$$
$$\partial_\lambda F_{\mu\nu} + \partial_\mu F_{\nu\lambda} + \partial_\nu F_{\lambda\mu} = 0 \qquad (1.1)$$

plus the **Lorentz force law**

$$\frac{dp_\mu}{dt} = qF_{\mu\nu}\frac{dx^\nu}{dt}, \qquad (1.2)$$

where $\mu, \nu, \lambda \in \{0,1,2,3\}$ and μ_0 and q are constants.

Maxwell's equations tell us how the electromagnetic field reacts to the presence and flow of electrically charged objects. We describe the electromagnetic field using the **electromagnetic field tensor** $F_{\mu\nu}$ and electric charges using the **current density** J_μ.[6] The electromagnetic field tensor is defined in terms of the **electromagnetic potential** A_μ as follows:

$$F_{\mu\nu} = \partial_\mu A_\nu - \partial_\nu A_\mu. \qquad (1.3)$$

In turn, the Lorentz force law tells us how charged objects are influenced by the electromagnetic field.

[6] A tensor is an object with more than one index and in this sense is a generalization of vectors. It is also possible to call vectors 1−tensors since they have exactly one index. We talk about this in more detail in Appendix A.1.

To summarize:

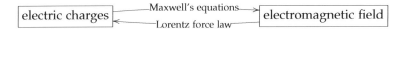

[7] What we do here is simply to give special names to certain components of the electromagnetic field tensor $F_{\mu\nu}$. Since, $\mu, \nu \in \{0,1,2,3\}$ the field tensor can be written as a (4×4) matrix:

$$F_{\mu\nu} = \begin{pmatrix} F_{00} & F_{01} & F_{02} & F_{03} \\ F_{10} & F_{11} & F_{12} & F_{13} \\ F_{20} & F_{21} & F_{22} & F_{23} \\ F_{30} & F_{31} & F_{32} & F_{33} \end{pmatrix}$$
$$= \begin{pmatrix} 0 & -E_1/c & -E_2/c & -E_3/c \\ E_1/c & 0 & -B_3 & B_2 \\ E_2/c & B_3 & 0 & -B_1 \\ E_3/c & -B_2 & B_1 & 0 \end{pmatrix}$$

There are different ways to write these equations. The form given here is especially useful since it is compact and useful for fundamental considerations. For many real world applications it's better to "unpack" these equations by introducing the electric field $\vec{E}$ and the magnetic field $\vec{B}$:[7]

$$E_i \equiv cF_{i0}$$
$$B_i \equiv -\frac{1}{2}\epsilon_{ijk}F^{jk},$$

where $i, j \in \{1,2,3\}$ and ϵ_{ijk} denotes the Levi-Civita symbol.

We can write the field tensor like this because it is antisymmetric: $F_{\mu\nu} = -F_{\nu\mu}$ (or in matrix form $F^T_{\mu\nu} = -F_{\mu\nu}$). We will derive later why this is the case. The antisymmetry implies that there must be zeroes on the diagonal since, for example, $F_{11} = -F_{11}$, which is only true for $F_{11} = 0$. Moreover, we have, for example, $F_{12} = -F_{21}$ and therefore there are in total only 6 independent components.

Moreover, the zeroth component of the electromagnetic current J_μ gets a special name: $J_0 \equiv c\rho$, where c denotes the speed of light and ρ the **charge density**.[8]

[8] We will discuss in Chapter 8 in detail why the charge density, in some sense, also can be thought of as some kind of current. The main idea will be that while the spatial components J_i describe how charge flows between different locations, ρ describes the flow between different points in time at a fixed location.

The remaining three components J_i with $i \in \{1,2,3\}$ are then called the **electric current density** and usually written as an ordinary three-vector $\vec{J}$.[9]

[9] A "three-vector" is a vector with three components while a "four-vector" is a vector with, well, four components. Four vectors are what we usually use in special relativity because here, spatial $i \in \{1,2,3\}$ components and temporal component 0 are mixed. This is also true here since a charge at rest ($\rho \neq 0, \vec{J} = 0$) is described by a different observer who moves relative to the first observer as a moving charge ($\rho \neq 0, \vec{J} \neq 0$). This is discussed in more detail in Chapter 6.

With these definitions, Maxwell's equations read:[10]

$$\nabla \cdot \vec{E} = \frac{\rho}{\epsilon_0}$$
$$\nabla \times \vec{B} - \mu_0 \epsilon_0 \frac{\partial \vec{E}}{\partial t} = \mu_0 \vec{J}$$
$$\nabla \cdot \vec{B} = 0$$
$$\nabla \times \vec{E} + \frac{\partial \vec{B}}{\partial t} = 0 \quad (1.4)$$

and the Lorentz force law reads:

$$\frac{d\vec{p}}{dt} = q\left(\vec{E} + \frac{d\vec{x}}{dt} \times \vec{B}\right). \quad (1.5)$$

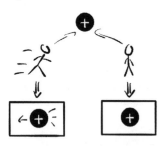

[10] This is shown explicitly in Appendix A.16.2. But be warned that rewriting Maxwell's equations in terms of $\vec{B}$ and $\vec{E}$ is far from trivial and it doesn't make much sense to spend much thought on this derivation at this point.

Of course, it's possible to unpack these equations even further, for example, by writing a separate equation for each component. This is, in fact, what Maxwell did historically. But this is not very useful and doesn't lead to further insights.

Now, with these equations at hand (in whatever form) the main task in electrodynamics is to solve them for different boundary conditions. As a result, we get formulas which describe the magnetic and electric field configuration in a given system. These formulas tell us that the electric and magnetic field configurations look, for example, like this:

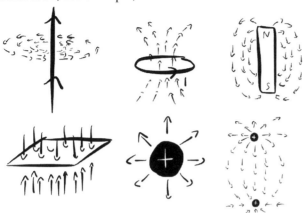

Moreover, we can then use the Lorentz force law to derive how charged objects react to these field configurations.

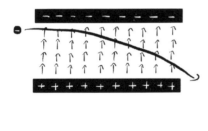

One of the most important phenomena described by Maxwell's equations are electromagnetic waves. An electromagnetic wave is a wave-like electric and magnetic field configuration which is able to travel through space.

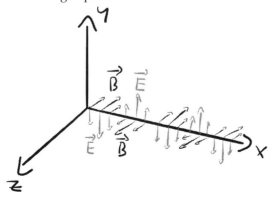

This is possible because a changing magnetic field leads to a changing electric field and, in turn, a changing electric field leads to a changing magnetic field.[11]

$$\text{changing } \vec{B} \longrightarrow \text{changing } \vec{E} \longrightarrow \text{changing } \vec{B} \longrightarrow \ldots$$

To be a bit more precise: using Maxwell's equations we can derive the so-called **wave equations**

$$\nabla^2 \vec{E} = \mu_0 \epsilon_0 \frac{\partial^2}{\partial t^2} \vec{E}$$

$$\nabla^2 \vec{B} = \mu_0 \epsilon_0 \frac{\partial^2}{\partial t^2} \vec{B}.$$

Solutions of these equations describe wave-like electric and magnetic field configurations ($\vec{E} \propto \cos(\omega t - kx)\vec{e}_y$).[12]

After this certainly overwhelming first glance at what electrodynamics is all about, let's take a step back and talk about the concepts introduced here in a bit more detail.[13]

[11] This fact is encoded in Maxwell's equations. For example, the fourth line in Eq. 1.4:

$$\nabla \times \vec{E} + \frac{\partial \vec{B}}{\partial t} = 0$$

tells us whenever our magnetic field is changing in time ($\frac{\partial \vec{B}}{\partial t} \neq 0$), there will be a non-zero electric field strength $\vec{E}$.

[12] Take note that it is possible to find special wave-like solutions which do not travel anywhere. These are known as standing wave solutions and can be understood as linear combinations of the more fundamental plane wave solutions. We will discuss this in detail in Section 5. Also note that we could equally write here $\vec{E} \propto \cos(kx - \omega t)\vec{e}_y$ since the cosine is an even function and hence it makes no difference if we reverse its argument.

[13] As mentioned above, each concept will be discussed several times from multiple perspectives.

Part I
What Everybody Ought to Know About Electrodynamics

"Only connect! That was the whole of her sermon ... Live in fragments no longer."

E. M. Forster

PS: You can discuss the content of Part I with other readers and give feedback at www.nononsensebooks.com/edyn/part1.

In the following two chapters, we will focus solely on the fundamental concepts and equations of electrodynamics. The mathematical tools needed to describe electrodynamics are discussed in the Appendices and it's entirely up to you *when* you read them. This is really a matter of taste and you can read this book however you like.

▷ For example, you can start by reading the appendices first to get a solid understanding of the required tools.

▷ Alternatively, you can read the corresponding appendix whenever a given tool shows up for the first time in the text.[14]

[14] The relevant appendix is always mentioned in a sidenote like this.

▷ But my personal recommendation would be that you start by reading Chapter 2 and Chapter 3 quickly without jumping around. During this first run, you can simply write down a little comment or question mark in the margin every time a mathematical concept is used that you're not familiar with.[15] Then, once you've finished reading these two chapters, you can go back to these question marks and read the corresponding appendices to deepen your understanding.

[15] That's why the margin is so big.

2

Fundamental Concepts

In this chapter, we discuss the basic notions needed to describe electrodynamics. This establishes the language we will use in the rest of the book.

We will start with the most basic notion: electric charge. Afterwards, we will discuss how we can describe electric charges in the context of electrodynamics using the concepts: charge density, electric current and electric current density. While the charge density is what we use to describe the locations of electric charges, the electric current and current density are used to describe how these charges move around.

We will then talk about the various concepts we can use to describe how electric charges influence each other, including the electromagnetic field, the electric field, the magnetic field and the electromagnetic potential. Each of these concepts have strengths and weaknesses depending on the task at hand. For example, the electric field and the magnetic field are helpful to analyze the electromagnetic field in specific real-world situations. However, the common origin of the electric field and the magnetic field becomes especially transparent when we introduce the electromagnetic potential.[1]

[1] The electric and magnetic fields can be calculated directly once the electromagnetic potential is specified.

Now, as promised, let's start with the simplest but arguably most fundamental notion in electrodynamics: Electric charge.

2.1 Electric charge

Each elementary particle is characterized by certain labels like its mass. Since these labels determine how particles behave in experiments, we use them to define and distinguish different types of elementary particles.

One of the most famous labels is what we call **electric charge**. But there are also others like the weak charge ("isospin"), the strong charge ("color") or the spin of the particle.

For example, an electron is characterized by the labels:

▷ mass: $9,109 \cdot 10^{-31}$ kg,

▷ spin: $\frac{1}{2}$,

▷ electric charge: $1,602 \cdot 10^{-19}$ C,

▷ weak charge, called weak isospin: $-\frac{1}{2}$,

▷ strong charge, called color charge: 0.

[2] One of the most beautiful aspects of modern physics is how we can understand the origin of all these labels using symmetries. We will not discuss this in detail here but if you're interested you may enjoy

Jakob Schwichtenberg. *Physics from Symmetry*. Springer, Cham, Switzerland, 2018a. ISBN 978-3319666303

An important thing these labels have in common is that they determine how (if at all) a particle takes part in fundamental interactions.[2] For example, a particle without electric charge does not take part in electromagnetic interactions. Analogously, only particles which carry a non-zero strong charge take part in strong interactions.

The way a particle takes part in a given interaction depends on the actual value of the charge. For example, a particle with **negative electric charge** (e.g. an electron) behaves different than a particle with **positive electric charge** (e.g. a proton). Similarly, a particle with a large electric charge behaves differently than a particle with a smaller charge. A crucial point is that not only different values for the electric charge are possible, but also different signs. This means that the charges of different particles can cancel and even yield zero when an object carries equal amounts of positive and negative charge. Another important property is that like charges repel and unlike charges attract.

Since our macroscopic everyday objects consist of elementary particles, we have to use the same labels to describe them. The electric charge of a macroscopic object is simply the sum of the charges of all elementary particles. An extremely surprising fact of nature is that the overall electric charge of almost all macroscopic objects is exactly zero even though each of them consists of billions of charged elementary particles (electrons, protons).[3] In other words, the charges of the billions of elementary particles inside any macroscopic object almost always average out to exactly zero. This comes about since protons have exactly the same charge as an electron, only with a different sign.[4]

Another important aspect of electric charge is that it's **conserved**. In other words, the total amount of electric charge in any closed system always stays the same.[5] This means that if the charge of an object is changed, the charge of a different object must also change by an equal and opposite amount so that the total amount of charge remains constant.[6]

[3] Take note that protons are not really elementary particles since they consist of three more fundamental quarks.

[4] This is known as the **quantization of electric charge** and is a completely non-trivial observation. So far, there is no generally accepted explanation for this curious fact of nature. In principle, the charges of protons and electrons could be completely arbitrary numbers and there is no reason why they should have anything to do with each other.

[5] This can also be understood using symmetry arguments and Noether's theorem, c.f. [Schwichtenberg, 2018a]. However, at this point the best we can do is to collect these experimental facts and then see how we can build them into our theory.

[6] In the illustration below, the letter "C" stands for "Coulomb" which is the SI-unit of electric charge.

$$Q=0C \quad Q=0C \quad \} \quad Q_{total}=0C$$

$$Q=2C \quad Q=-2C \quad \} \quad Q_{total}=0C$$

Formulated differently, electric charge can never be produced or destroyed, only moved from one place to another. In macroscopic applications, we usually move electrons around. If we want to generate a positive charge for some object, we need to remove electrons from it. Of course, the electrons then have to go somewhere and, as a result, the place where they end up will be negatively charged.

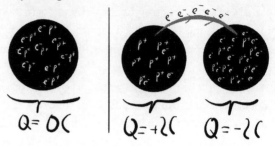

The conservation of electric charge holds even at the most fundamental level. All interactions of fundamental particles which have been observed so far conserve electric charge. For example, while an electron can indeed appear seemingly out of nowhere, at the same time a positron (the electrons' antiparticle) always appears which carries exactly the opposite charge. Therefore, the total electric charge before and after such a process remains exactly the same.

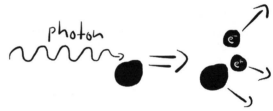

This also implies that electric charge is conserved *locally*. What this means is that electric charge doesn't vanish at one point and then magically appears at another place. This would be no problem if electric charge were only conserved *globally*. If only the total charge inside the whole universe would be conserved, there would be no problem if electric charge vanishes from a given small system since it could have materialized somewhere else. However, this is not the case. Electric charge is conserved no matter how small our system is, even for elementary particle processes.[7] Formulated differently, no matter how small the volume is we are considering, the total charge inside remains the same (if the system is closed).

Now, in interesting systems we are usually dealing with more than one charged object and want to describe how these move.

[7] Take note that local conservation implies global conservation, but not vice versa.

In particular, the total charge of any macroscopic charged object is the result of the individual charges of an incredibly large number of electrons and protons. To keep track of the locations and movements of a large number of charged objects, we need new tools. These tools are the topic of the following sections.

2.2 Charge density

The first extremely useful notion to describe systems with lots of charged objects is **charge density**.[8] Charge density $\rho(\vec{x}, t)$ is defined as "charge per unit volume" and enables us to keep track of the locations of charges in our system.

[8] An alternative name for the charge density is **charge distribution**.

While electric charge is quantized and therefore only appears in discrete lumps, in macroscopic systems we are usually dealing with *continuous* charge densities. This is a valid approach since any macroscopic region contains an incredibly large number of fundamental charges and the discrete nature of the electric charge makes no difference. If we could zoom in until we "see" individual electrons, we would notice that the charge density varies wildly from point to point. However, in macroscopic applications, we are so far away and always consider systems which consist of lots of elementary charges such that the charge distribution appears smooth.

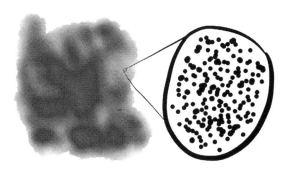

The main idea behind the charge density is that we can describe this incredibly large number of fundamental charges sufficiently accurately by taking the total charge within some manageable region and dividing it by the region's volume.

The **total charge** inside any volume V is then given by the integral over the charge density:

$$Q = \int_V d^3x \, \rho(\vec{x}, t). \tag{2.1}$$

So when we integrate the charge density $\rho(\vec{x}, t)$ over a specific region V of our system and end up with a large number Q, we know immediately that there is a large amount of electric charge concentrated in this region.

$$\int_V dV = Q \gg q = \int_V dV$$

In other words, the charge density encodes where electric charges are concentrated in our system. Usually, our charge density is a smooth function which we can think of as arising from averaging over many individual charges. For example, when we have a ball with total charge Q which is evenly distributed through the ball, the charge density is constant:

$$\rho = \frac{Q}{V}, \tag{2.2}$$

where V is the volume of the ball. Outside the ball, the charge density is zero.

Of course, we can also use the charge density if there is only one charged object in our system. In this case, the charge density is zero everywhere except at one specific point. Any integral over a region which contains the location of the object, yields simply the charge of the object.

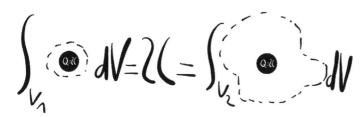

The correct mathematical tool to describe such a charge density is the delta distribution.[9] Specifically, the charge density of a system with just one charged object located at $\vec{x}_0$ is

[9] If you are unfamiliar with the delta distribution, have a look at Appendix C.

$$\rho(\vec{x}) = q\delta(\vec{x} - \vec{x}_0), \qquad (2.3)$$

where q is the charge of the object. Any integral over a region V_0 which contains $\vec{x}_0$ yields exactly q, as it should be

$$\begin{aligned}\int_{V_0} d^3x\, \rho(\vec{x},t) &= \int_{V_0} d^3x\, q\delta(\vec{x}-\vec{x}_0) \\ &= q \int_{V_0} d^3x\, \delta(\vec{x}-\vec{x}_0) \\ &= q\,. \end{aligned} \qquad \circlearrowleft \int dx\, \delta(x) = 1 \qquad (2.4)$$

Usually, we not only want to describe the locations of charged objects, but also how they move. This is possible by using what we call current and current density. These concepts are what we will talk about next.

2.3 The electric current

In general, a current describes how a given quantity (heat, electric charge, etc.) moves through a system. There can be moving objects everywhere. Therefore, a current encodes information about the flow of the quantity at each point in space. In other words, a current $I(\vec{x})$ is a function that eats a point in space $\vec{x}$ and spits out a number $I(\vec{x})$. In electrodynamics this number tells us the amount of charge that passes through the given location per unit time

$$I = \frac{dQ}{dt}\,. \qquad (2.5)$$

For simplicity, let's assume we are dealing with a one-dimensional system, for example, an extremely thin wire. Moreover, let's assume that the charge density in the wire is constant ρ_c and that our charges move with the constant velocity v_c.

During a given time interval Δt, the distance our charges travel in a line segment is $\Delta L = v_c \Delta t$. The total charge contained in a line segment is $\Delta Q = \rho_c \Delta L$. This means that the total charge contained in this line segment is

$$\Delta Q = \rho_c v_c \Delta t. \tag{2.6}$$

A crucial observation is that the total amount of electric charge passing any specific point P during this time interval is equal to the amount of charge contained in this line segment:

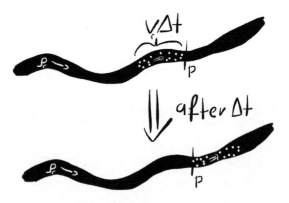

The electric current inside the wire is therefore

$$I = \frac{\Delta Q}{\Delta t} = \rho_c v_c. \tag{2.7}$$

In general, of course, our charges can move in more than just one dimension. If this is the case, we need to keep track of the various directions in which our charges move. Mathematically, this means that a simple number at each location is no longer sufficient and we need vectors instead. The correct tool to describe the flow of charges in three dimensions is called current density. A current density yields a vector at each point in space.[10] The direction of the vector at a given point describes the direction of the flow. The length of the vector describes how much is flowing.

[10] Mathematically, we use a vector field. We will talk more about fields in the next section and also in Appendix A.4.

Keeping track of charges in three dimensions can easily become quite messy. Therefore, we first discuss the concept "current density" in the context of a simple special situation. Afterwards, we will discuss current densities in general.

Analogous to what we did above, we consider again a constant charge density ρ_c and assume that all charges move with the velocity $\vec{v}_c$ in a common direction.[11] In one dimension, our task was to describe how many charges pass any given point per given time interval. Now, in three dimensions our task is to describe how many charges pass any given frame per time interval.[12]

[11] We can imagine, for example, that $\vec{v}_c$ describes the average velocity.

[12] Take note that the frame is not necessarily at right angles to the direction of flow of the charges.

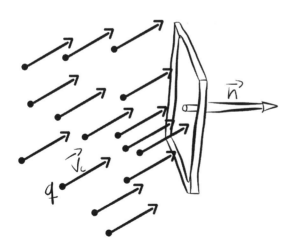

How many of the charges shown here will pass the frame during the interval Δt?

Analogous to how we were able to answer this question for

the one-dimensional problem using a line segment, we now consider a "segment" of our three-dimensional space. Here our "segment" is a prism with a base given by the frame and an edge of length $|\vec{v}_c|\Delta t$, which is again the distance our charges travel during the interval Δt.[13]

[13] Charges outside this prism are either too far away to reach the frame during the interval Δt or miss the frame.

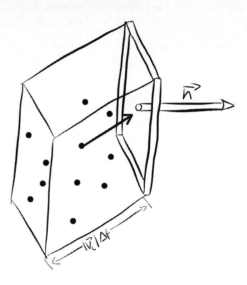

The number of charges which pass our frame are exactly those contained in the prism: $\Delta Q = \rho_0 V_{\text{prism}}$.[14] As usual, the volume of the prism is given by the product of its base and height. The height of the prism is $|\vec{v}_c|\Delta t \cos\theta$:

[14] This is completely analogous to what we discussed above for one-dimensional systems where exactly those charges contained in the line segment $|\vec{v}_c|\Delta t$ passed the point P.

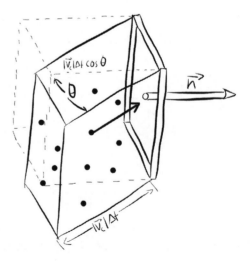

Therefore, if we denote the surface area of our frame with A, we can write the volume of the prism as

$$V_{\text{prism}} = A|\vec{v}_c|\Delta t \cos\theta. \qquad (2.8)$$

The total amount of electric charge passing through the frame is then

$$\Delta Q = \rho_0 V_{\text{prism}} = \rho_0 A |\vec{v}_c| \Delta t \cos\theta. \qquad (2.9)$$

By introducing a vector $\vec{n}$ of unit length normal to the frame and recalling the definition of the scalar product, we can write this formula as

$$\Delta Q = \rho_0 A \Delta t \vec{v}_c \cdot \vec{n}, \qquad (2.10)$$

since $\vec{v}_c \cdot \vec{n} = |\vec{v}_c||\vec{n}|\cos\theta = |\vec{v}_c|\cos\theta$, where θ is the angle between the two vectors.[15]

[15] The scalar product is discussed in Appendix A.2.

Using this equation, we can conclude that the amount of charge passing our frame per given time interval (= our electric current) is[16]

$$I \equiv \frac{\Delta Q}{\Delta t} = \rho_0 A \vec{v}_c \cdot \vec{n}. \qquad (2.11)$$

[16] This equation is completely analogous to Eq. 2.7.

An important point is that this formula yields a specific number (the electric current) for each specific frame $A\vec{n}$. Usually, it is more convenient to use a quantity which encodes information about the flow for any given frame and not only one specific one. For this purpose, we define the **electric current density** as

$$\vec{J} \equiv \rho_0 \vec{v}_c. \qquad (2.12)$$

This current density assigns a vector to each point in space, which is exactly the idea discussed above. Moreover, to get the current flowing through any specific frame, we simply have to multiply the current density by $\vec{n}A$:

$$\vec{J} \cdot \vec{n} A = \rho_0 \vec{v}_c \cdot \vec{n} A = I. \qquad (2.13)$$

In general, the magnitude of the current density $|\vec{J}|$ at one specific point describes the amount of electric charge which passes per unit time through an *infinitesimal* surface element which is at right angles to the direction of the flow. The direction of the charge density vector $\vec{J}$ encodes where the charges are flowing.

If we want to know how much electric charge is flowing through a more complicated surface S, we have to calculate the contribution from each infinitesimal element $\vec{dS}$ that the surface consists of and then calculate the sum of these contributions:

$$I = \int_S \vec{J} \cdot \vec{dS}. \tag{2.14}$$

The total amount of charge passing through the surface S during some time interval Δt is then given by

$$\Delta Q = \Delta t \int_S \vec{J} \cdot \vec{dS}. \tag{2.15}$$

Mathematically, we call $\int_S \vec{J} \cdot \vec{dS}$ a **surface integral** and it describes the **flux** of electric charge through the surface S.[17]

[17] To learn more about surface integrals and flux, read Appendix A.9.

So far, we have only talked about how we can describe the locations and the flow of charged objects. However, in electrodynamics we are primarily interested in how charged objects influence each other. To describe this, we need further mathematical tools and concepts and this is what will we talk about next.

2.4 The electromagnetic field

The question we want to answer in this section is: Which mathematical tool is the right one to describe electromagnetic interactions?

A first crucial hint is that electrically charged objects can influence each other even though they do not touch.

This means that electromagnetic interactions have to be mediated by *something*. We call this something the electromagnetic field and, as the name already indicates, mathematically we describe it using a field.[18] A **scalar field** is certainly not sufficient since charged particles get pushed through electromagnetic interactions in a certain direction, analogous to how a piece of paper gets pushed by wind.[19] Directional information are described using a vector at each point in space.

[18] If you're unfamiliar with "fields" and the different kinds of fields: scalar fields, vector fields, tensor fields, have a look at Appendix A.4.

[19] As mentioned in Appendix A.1, we describe wind using a vector field.

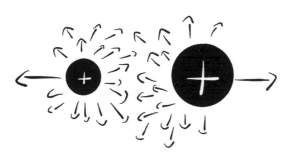

An important feature of the electromagnetic field is that even a **vector field** is not sufficient and we need instead a **tensor field**. The electromagnetic field tensor is an antisymmetric (4×4) matrix[20]

$$F_{\mu\nu}(t,\vec{x}) = \begin{pmatrix} 0 & F_{01}(t,\vec{x}) & F_{02}(t,\vec{x}) & F_{03}(t,\vec{x}) \\ -F_{01}(t,\vec{x}) & 0 & F_{12}(t,\vec{x}) & F_{13}(t,\vec{x}) \\ -F_{02}(t,\vec{x}) & -F_{12}(t,\vec{x}) & 0 & F_{23}(t,\vec{x}) \\ -F_{03}(t,\vec{x}) & -F_{13}(t,\vec{x}) & -F_{23}(t,\vec{x}) & 0 \end{pmatrix}$$

(2.16)

[20] The antisymmetry follows from the definition of the field tensor in terms of the electromagnetic potential A_μ, which is something we will discuss in the next section.

This tensor field, in some sense, assigns exactly two vectors to each point in space $\vec{x}$ at each point in time t. We need two vectors at each point to describe the electromagnetic field completely. The vectors at each location represent the strength and direction of the electromagnetic field.[21]

[21] This will make a lot more sense in a moment.

The most famous way to visualize this is to use a magnet and lots of tiny pieces of metal. If we put these pieces of metal around the magnet, they arrange in a specific pattern, which makes the electromagnetic field at least somewhat visible:

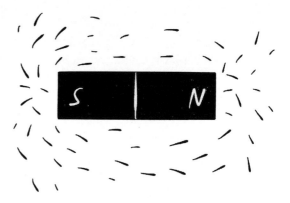

It is conventional to introduce the electric vector field $\vec{E}(t,\vec{x})$ and the magnetic vector field $\vec{B}(t,\vec{x})$ and work with them, instead of with the electromagnetic tensor field $F_{\mu\nu}(t,\vec{x})$.[22]

[22] Here c denotes the speed of light which is arguably the most important constant in electrodynamics. We will talk about it in detail in Chapter 6.

$$F_{\mu\nu}(t,\vec{x}) = \begin{pmatrix} 0 & F_{01}(t,\vec{x}) & F_{02}(t,\vec{x}) & F_{03}(t,\vec{x}) \\ -F_{01}(t,\vec{x}) & 0 & F_{12}(t,\vec{x}) & F_{13}(t,\vec{x}) \\ -F_{02}(t,\vec{x}) & -F_{12}(t,\vec{x}) & 0 & F_{23}(t,\vec{x}) \\ -F_{03}(t,\vec{x}) & -F_{13}(t,\vec{x}) & -F_{23}(t,\vec{x}) & 0 \end{pmatrix}$$

$$\equiv \begin{pmatrix} 0 & -E_1(t,\vec{x})/c & -E_2(t,\vec{x})/c & -E_3(t,\vec{x})/c \\ E_1(t,\vec{x})/c & 0 & -B_3(t,\vec{x}) & B_2(t,\vec{x}) \\ E_2(t,\vec{x})/c & B_3(t,\vec{x}) & 0 & -B_1(t,\vec{x}) \\ E_3(t,\vec{x})/c & -B_2(t,\vec{x}) & B_1(t,\vec{x}) & 0 \end{pmatrix}.$$

(2.17)

[23] Writing out explicitly that our fields depend on the location $\vec{x}$ and the time t makes our equations quickly overcluttered. Hence, usually we do not write these dependencies explicitly and only write $\vec{E}$ or $\vec{B}$. However, it is always important to keep in mind that, in general, these fields assign a different vector to each point in space and time. Take note that the superscript T here denotes transposition which is a transformation that turns a column vector into a row vector and vice versa. Here, I've used transposed vectors because column vectors don't fit nicely into a text paragraph.

Each of these two fields $\vec{E}(t,\vec{x}) = (E_1(t,\vec{x}), E_2(t,\vec{x}), E_3(t,\vec{x}))^T$ and $\vec{B}(t,\vec{x}) = (B_1(t,\vec{x}), B_2(t,\vec{x}), B_3(t,\vec{x}))^T$ assigns a vector to each point in space $\vec{x}$ at each point in time t.[23]

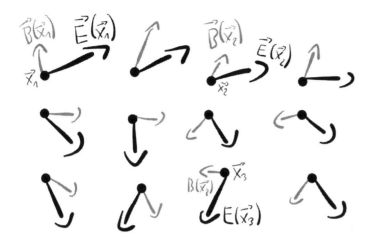

Now, what's the meaning of these little vectors at each point in space?

To understand this, let's recall what similar vectors describe in physics.[24] For example, for the vector field describing air, the little arrows attached to each point in space represent the motion of the individual air molecules. In addition, the electric current density $\vec{J}(t, \vec{x})$ also assigns a little vector to each point in space and is therefore also a vector field. Here the little arrows tell us the speed of our charges and in which direction they are moving.

Unfortunately, the interpretation of the electric field $\vec{E}(t, \vec{x})$ and the magnetic field $\vec{B}(t, \vec{x})$ is not so simple. For these fields the little arrows have a more abstract meaning since nothing is really flowing around.[25] Instead, the little arrows encode information about the somewhat abstract "physical field", which we can understand as follows.

The *direction* of the vector at a given location encodes in which direction a **test charge** would move if it were placed here.[26] The *magnitude* of the vector encodes how fast the test charge would accelerate as a result of, for example, the electric field. In this sense, these more abstract fields encode how something would flow *if* it were there.

[24] Again: If you're unfamiliar with vector fields, have a look at Appendix A.4.

[25] Historically, physicists did imagine that the electromagnetic field is a real substance like water and this substance called ether really does flow around. However, this "ether hypothesis" has been abandoned in modern physics since experiments never found any of the predicted ether effects.

[26] A test charge is an object with a tiny electric charge which we use to gather information about the electromagnetic field. Each charge has a direct influence on the electromagnetic field. But we want to use the test charge to really test what the electric field would look like in the absence of our "measuring device" and therefore want to minimize the influence of the object we are using as much as possible. Formulated differently, by using a tiny test charge we minimize the modification of the electric field due to our measurement procedure.

In practice the electric field strength at a point is measured by placing a small positive test charge at that point and measuring the force on it:

$$\vec{E}(\vec{x}) \equiv \frac{\vec{F}(\vec{x})}{q}, \qquad (2.18)$$

where q denotes the charge of the object we put at the location $\vec{x}$.[27]

[27] Analogously, the magnetic field at a point can be measured by using a *moving* test charge. Take note that we also use electric test charges to probe the magnetic field since, so far, no magnetic monopole has ever been observed. We will talk more about this point below.

This way we can deduce field configurations which look, for example, like this:

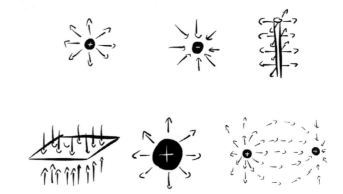

This whole physical field idea is somewhat abstract and certainly needs some time to get used to. However, the electric and the magnetic fields lose much of the mystery surrounding them as soon as we use them to describe specific situations. At this point, a possibly helpful point of view is that these abstract fields are simply useful mathematical devices to help us describe electromagnetic interactions. They assign little arrows to each point in space that encode how a charge would react if it were there, even if no charge is there. Maybe it helps that even Nobel laureate Richard Feynman had a somewhat blurry picture of the electromagnetic field in mind:[28]

[28] [Feynman, 2011]

I'll tell you what I see. I see some kind of vague showy, wiggling lines - here and there an E and a B written on them somehow, and perhaps some of the lines have arrows on them — an arrow here or there which disappears when I look too closely at it. When I talk about the fields swishing through space, I have a terrible confusion between the symbols I use to describe the objects and the objects themselves. I cannot really make a picture that is even nearly like the true waves. So if you have difficulty making such a picture, you should not be worried that your difficulty is unusual.

Before we move on, three short comments:

▷ Firstly, while thinking about the electromagnetic field in terms of the electric field and the magnetic field can be useful, for fundamental consideration this splitting of the electromagnetic field $F_{\mu\nu}$ into two seemingly separate parts $(\vec{E}, \vec{B})$ can be quite confusing.[29] Instead, the deep structure of electrodynamics becomes a lot more transparent as soon as we talk about the electromagnetic potential, which is the topic of the next section.

▷ Secondly, there is only one electromagnetic field in the universe and not many different ones. But it can be useful to imagine that each charge generates its own electromagnetic field and the total electromagnetic field is then the sum of all these hypothetical individual electromagnetic fields.[30] Alternatively, we can imagine that each charge generates a specific configuration of the surrounding electromagnetic field.[31]

▷ Thirdly, it is important to keep in mind that usually things in nature are subject to constant change. This means that our electromagnetic field, or analogously our electric and magnetic field, vary over time. Therefore, the little vectors attached to each point in space change their direction and magnitude as time passes.

[29] An important point we will discuss later is, for example, that different observers do not necessarily agree on what they call electric and what they call magnetic field. The electric field of one observer may appear, at least in part, like a magnetic field for another observer. This is discussed in Chapter 6.

[30] This is possible because the Maxwell equations are linear and we can therefore generate new solutions by using **superpositions** of known solutions. We will discuss this in detail in Chapter 4.

[31] In a similar spirit, the modern point of view in quantum field theory is that, for example, there is only one electron field in the whole universe. Each electron that we observe is simply an excitation of this electron field and this explains why every electron we have observed so far has exactly the same properties (mass, charge, etc.).

Next, as promised, let's talk about the electromagnetic potential.

2.5 The electromagnetic potential

Due to their appearance in Newtonian mechanics, **potentials** are a concept that should be much more familiar than fields.

A potential tells us how much potential energy a given object has if we release it at a specific location. For example, if we release an object of mass m in the gravitational potential $\phi(x)$ at the location x_1, we know that it has potential energy $m\phi(x_1)$. If we release the object it's possible that it will start moving. This means that its kinetic energy changes from zero to a non-zero value and becomes larger. Since energy must be conserved, the kinetic energy must have come from somewhere. This somewhere is the corresponding potential.

[32] We discuss scalar fields in Appendix A.1.

Usually, in mathematical terms a potential is a scalar field, in the sense that it eats a location and spits out a number.[32] Formulated differently, a potential usually assigns to each point in space a number and not an arrow (or tensor).

[33] However, take note that not every force is the result of a potential.

Moreover, there is a close relationship between potentials and the forces that act on an object.[33] For a potential ϕ, the corresponding force $\vec{F}$ that acts on an object is[34]

[34] The symbol ∇ is known as "del" or "nabla" and $\nabla \phi$ is known as the **gradient** of the scalar field ϕ. Gradients are discussed in detail in Appendix A.10.

$$\vec{F} = -\vec{\nabla}\phi. \qquad (2.19)$$

Therefore, given a potential, we can immediately calculate its effect on an object using Newton's second law $\vec{F} = m\vec{a}$.

Completely analogous to the gravitational potential, we can also define an **electromagnetic potential**.

This potential is a convenient tool to describe electromagnetic interactions since once the potential is defined, we can use the relationship between forces and potentials to calculate how charges move in a non-zero potential.[35]

The crucial point is that in order to describe electromagnetic interactions, a scalar potential is not sufficient. Instead, we need a four-vector potential $A_\mu(t, \vec{x})$, where $\mu = 0, 1, 2, 3$. We will discuss the meaning of this vector potential in detail in Chapter 8.

For the moment, we only note that the electromagnetic potential is characterized by 4 numbers at each point in space and time $A_\mu(t, \vec{x}) = (A_0(t, \vec{x}), A_1(t, \vec{x}), A_2(t, \vec{x}), A_3(t, \vec{x}))^T$ and that the electric and magnetic fields can be calculated immediately once the electromagnetic potential is specified:[36]

$$E_i = c(\partial_i A_0 - \partial_0 A_i)$$
$$B_i = \epsilon_{ijk} \partial_j A_k, \quad (2.20)$$

where $i, j, k \in \{1, 2, 3\}$. We can also write this as a vector equation[37]

$$\vec{E} = c\nabla A_0 - \partial_t \vec{A}$$
$$\vec{B} = \nabla \times \vec{A}. \quad (2.21)$$

We can see that the electric and magnetic fields have indeed a common origin, namely the electromagnetic potential A_μ. In this sense it makes sense to talk about the *electromagnetic* potential.

Working with the electromagnetic potential is not only conceptually attractive, but also more economical. The magnetic field $\vec{B}$ and the electric field $\vec{E}$ have 3 components each. This means

---

[35] Take note that potentials cannot be directly measured. We can see this, because the relationship between a potential and the corresponding force (which is something we can actually measure) is somewhat indirect (Eq. 2.19). In particular, take note that we can always add an arbitrary constant to the potential without changing the resulting force:

$$\phi \to \phi' = \phi + C$$
$$\Rightarrow \vec{F} \to -\nabla \phi' = -\nabla(\phi + C)$$
$$= -\nabla \phi - \underbrace{\nabla C}_{=0}$$
$$= -\nabla \phi = \vec{F} \checkmark$$

This freedom is usually called "gauge freedom" and we will talk more about it in Chapter 7.
In contrast, we can actually measure the gravitational field, the electric field $\vec{E}$ and magnetic field $\vec{B}$ since the relationship between the field and the corresponding force is a lot more direct (Eq. 2.18). In other words, as mentioned above, we can measure the field strength and direction at a particular location by observing the resulting force acting on a test charge.

[36] As already mentioned above, the little superscript "T" denotes transposition, which means that we turn a row vector into a column vector and vice versa. This is useful since writing A_μ as a column vector would destroy the layout of the page completely.

[37] It is often more convenient to write equations in index notation. The cross product $\vec{A} \times \vec{B}$ reads in index notation $\epsilon_{ijk} A_i B_j$, where ϵ_{ijk} is the Levi-Civita symbol.

that if we describe electrodynamics in terms of $\vec{B}$ and $\vec{E}$, we need to keep track of 6 functions $(E_1(t,\vec{x}), E_2(t,\vec{x}), \ldots)$.

In contrast, if we work with A_μ, we only have to deal with 4 functions $(A_0(t,\vec{x}), A_1(t,\vec{x}), A_2(t,\vec{x}), A_3(t,\vec{x}))$.[38]

[38] Take note that this is still not the most economical choice possible. In fundamental terms electromagnetic interactions are mediated by particles called photons. A photon has only 2 internal degrees of freedom and therefore, in principle, two functions are sufficient to describe it. However, this is far beyond the scope of this book. To describe photons properly, we need to use quantum field theory.

Moreover, since the electric and magnetic fields appear as components of the electromagnetic field tensor, we can also express the tensor field itself using the electromagnetic potential[39]

$$F_{\mu\nu} = \partial_\mu A_\nu - \partial_\nu A_\mu \qquad (2.22)$$

[39] The electric and magnetic fields were defined as components of the electromagnetic tensor in Eq. 2.17. The definition of the field strength tensor in terms of the potential may seem rather strange and ad hoc. However, we will see in Chapter 8 that this equation has a really intuitive meaning.

Before we move on and talk about the equations describing the interplay between the various concepts introduced in this chapter, let's summarize what we have learned so far.

2.6 Summary

There are different tools we can use to describe electromagnetic interactions. One economical possibility is the electromagnetic potential $A_\mu(t,\vec{x})$, which assigns four numbers to each point in space.[40] For many concrete applications it is conventional to

[40] $A_\mu = (A_0(t,\vec{x}), A_1(t,\vec{x}), A_2(t,\vec{x}), A_3(t,\vec{x}))^T$

use the electric field $\vec{E}(t,\vec{x})$ and magnetic field $\vec{B}(t,\vec{x})$ instead. Both can be calculated directly from the given electromagnetic potential (Eq. 2.20). A third possibility is to use the electromagnetic tensor $F_{\mu\nu}(t,\vec{x})$. This tensor is an antisymmetric (4×4) matrix which contains the components of the electric and magnetic fields as entries (Eq. 2.17). Moreover, the electromagnetic field tensor can be calculated directly using the electromagnetic potential (Eq. 2.22).

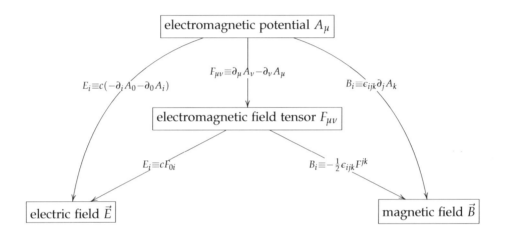

In addition to these concepts that deal with electromagnetic interactions directly, we also need notions to describe the objects that are influenced by them. The most basic notion in this context is electric charge. Only objects with a non-zero electric charge take part in electromagnetic interactions. Moreover, usually we are interested in systems that consist of lots of individual charges. To describe the locations of charges in such systems, we use charge density $\rho(t,\vec{x})$ which encodes exactly where charges are concentrated within our system. In addition, to describe how charged objects move, we use electric current I and electric current density $\vec{J}(t,\vec{x})$.[41]

Now that we have established these fundamental physical concepts, we can discuss the fundamental equations of electrodynamics.

[41] Recall that the flow of a quantity like electric charge is a directional quantity and hence, we need vectors to describe it completely.

3
Fundamental Equations

The most important equations of electrodynamics are Maxwell's equations (Eq. 1.4) and the Lorentz force law (Eq. 1.5).[1] The Lorentz force law describes how charged objects react to the presence of the electric and magnetic field. Maxwell's equations describe how non-zero electric and magnetic field strengths are generated by charges and currents. Moreover, they also describe how the electric and magnetic fields influence each other. This interplay is the reason why electromagnetic waves exist which are described by the so-called wave equations. In addition, there is another incredibly important equation which encodes the fact that electric charge is conserved: the continuity equation.[2] All this is summarized by the following diagram.

[1] Reminder: Eq. 1.4 reads
$$\nabla \cdot \vec{E} = \frac{\rho}{\epsilon_0}$$
$$\nabla \times \vec{B} - \mu_0 \epsilon_0 \frac{\partial \vec{E}}{\partial t} = \mu_0 \vec{J}$$
$$\nabla \cdot \vec{B} = 0$$
$$\nabla \times \vec{E} + \frac{\partial \vec{B}}{\partial t} = 0 \, .$$
and Eq. 1.5
$$\frac{d\vec{p}}{dt} = q \left(\vec{E} + \frac{d\vec{x}}{dt} \times \vec{B} \right) . \tag{3.1}$$

[2] The wave equations and the continuity equation can both be derived using Maxwell's equations.

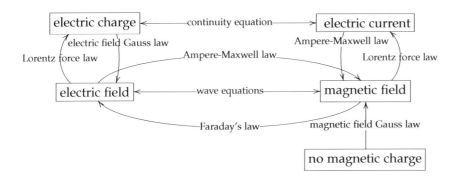

Whenever you feel lost in the following sections, come back here to see how what we are doing fits into the bigger picture. But there is no reason to spend much time looking at this diagram if you're just starting this chapter.

Take note that we will not discuss explicit applications of these equations in the following sections since it would be too easy to get lost. Applications are discussed separately in Part II.

With that said, let's start discussing the fundamental equations of electrodynamics in detail. As usual, we start with the simplest one.

3.1 The continuity equation

In plain language, the continuity equation says that when the amount of electric charge within some volume gets smaller, it must have been transported to a location somewhere outside the volume. Analogously, it says that when the electric charge gets larger within some volume, it must have come from somewhere outside the volume.[3]

[3] Take note that, of course, it's also possible to have a continuity equation for a quantity which is *not* conserved. In this case, there is an additional source term σ in the equation which describes how much of the quantity gets produced per unit time:

$$\frac{\partial \rho}{\partial t} = \sigma - \nabla \cdot \vec{j}.$$

The equation then tells us that when the amount of the quantity we want to describe gets larger, it must either come from somewhere else or be produced inside the volume. However, in electrodynamics, we want to describe electric charge, which is conserved and therefore there is no such source term.

Using the concepts introduced in the previous sections, we can formulate this statement directly in mathematical terms.

The amount of electric charge within some volume V is given by[4]

$$Q = \int_V \rho dV. \quad (3.2)$$

The change in the amount of electric charge within the volume during some time interval Δt is given by[5]

$$\Delta Q = -\Delta t \frac{\partial}{\partial t} \int_V \rho dV. \quad (3.3)$$

Here $\frac{\partial}{\partial t} \int_V \rho dV$ is the rate at which the amount of electric charge changes and if we multiply it by an interval Δt, we get the total change in the amount of electric charge.

Now, the whole idea of the continuity equation is that if ΔQ is non-zero, electric charge must have moved into the volume or, alternatively, has been removed from the volume because it got pushed across the boundary of the volume. Information about the movement of electric charge is encoded in the current density $\vec{J}$. In Section 2.3, we already discussed that the total amount of electric charge passing a given surface S during an interval Δt is given by (Eq. 2.15)[6]

$$\Delta \tilde{Q} = -\Delta t \oint_S \vec{J} \cdot d\vec{S}. \quad (3.4)$$

Since electric charge is conserved, we know that the change in the amount of electric charge inside the volume must be a result of electric charge passing the surface of the volume. Hence, charge conservation means that $\Delta Q = \Delta \tilde{Q}$ and therefore we find

$$\Delta Q = \Delta \tilde{Q}$$
$$\therefore \quad \Delta t \frac{\partial}{\partial t} \int_V \rho dV = -\Delta t \oint_S \vec{J} \cdot d\vec{S}. \quad (3.5)$$

Since the interval Δt appears on both sides it drops out from the equation and we can conclude.

$$\boxed{-\frac{\partial}{\partial t} \int_V \rho dV = \oint_S \vec{J} \cdot d\vec{S}} \quad (3.6)$$

This is the **integral form of the continuity equation**. It is important to take note that this equation really describes the *local*

[4] As a reminder: we use the symbol ρ to describe the charge density. If we integrate the charge density over some volume, we get the total amount of charge within this volume.

[5] Take note that we can't simply state, for example, that a positive ΔQ means that charge has moved into the volume. While it is positive that ΔQ is positive because, for example, protons moved into the volume, it is also possible that electrons left the volume. Since electrons are negatively charged, if they leave the volume the total charge inside it becomes more positive. In practice, it's almost always electrons which are moving around.

[6] The symbol $\oint_S$ indicates that we are integrating over a *closed* surface. Here, we have a closed surface since S is the surface of the volume V. The minus sign appears here as a result of the convention that we define the electric field in terms of its force on a *positive* test charge (c.f. Eq. 2.18). Moreover, we use the convention that the vector that characterizes our surface, $d\vec{S}$, is outward normal. Hence we consider an outward flux of positive charge. Feel free to ignore details like this on a first reading.

conservation of electric charge, since we can make the volume V as small as we please.[7]

[7] The distinction between local and global conservation of electric charge was discussed in Section 2.1.

In words, the continuity equations tells us:

> The rate at which the charge density changes in a given volume is exactly equal to the net amount of electric charge flowing through the surface of the volume per unit time.

An important observation is that we have on the left-hand side an integral over the *volume*, but on the right-hand side we have an integral over the *surface* of the volume. Luckily, there is a clever trick known as Gauss's theorem that allows us to get a volume integral on the right-hand side too. This, in turn, allows us to write the equation without the integral. The resulting equation is known as the *differential form* of the continuity equation. This differential form allows us to encode exactly the same idea in a more compact form.[8]

[8] The differential form is also a lot more useful for many advanced applications.

Here's how this works.

First of all, Gauss's theorem is a fundamental theorem of vector calculus which allows us to rewrite a surface integral as a volume integral:[9]

$$\oint_S \vec{J} \cdot d\vec{S} = \int_V \nabla \cdot \vec{J} dV. \tag{3.7}$$

[9] For a detailed discussion of Gauss's theorem, see Appendix A.13. An important point is that we get the volume V enclosed by the surface S on the right-hand side and not some arbitrary volume.

Here, $\nabla \cdot \vec{J}$ denotes the divergence of $\vec{J}$.[10] Gauss's theorem holds for *any* vector field and therefore also for our current density $\vec{J}$.

[10] The divergence of a vector field is discussed in Appendix A.11.

Using Gauss's theorem we can rewrite our continuity equation as follows

$$-\frac{\partial}{\partial t} \int_V \rho dV = \oint_S \vec{J} \cdot d\vec{S} \qquad \text{this is Eq. 3.6}$$

↧ Gauss's theorem

$$-\frac{\partial}{\partial t} \int_V \rho dV = \int_V \nabla \cdot \vec{J} dV$$

↧

$$\int_V \left(\frac{\partial}{\partial t} \rho + \nabla \cdot \vec{J} \right) dV = 0.$$

We can now use that we made no assumptions about the volume and therefore the only possibility to ensure that the integral vanishes is if the integrand $(\frac{\partial}{\partial t}\rho + \nabla \cdot \vec{j})$ is zero:

$$\boxed{\frac{\partial}{\partial t}\rho + \nabla \cdot \vec{j} = 0} \tag{3.8}$$

This is the **differential form of the continuity equation**.[11]

[11] We will see in a moment that, completely analogously, there is an integral form and a differential form of Maxwell's equations.

Now, after this short discussion of the only fundamental equation which deals exclusively with electric charges and their flow, it's time to move on to equations which describe the interplay between charges and the electromagnetic field. As already mentioned above, these equations are known as Maxwell's equations and the Lorentz force law. In the following sections, we will discuss them one by one.

3.2 The Lorentz force law

The Lorentz force law (Eq. 1.5) describes how electric charges react to the presence of a non-zero electric or magnetic field strength. It consists of two parts. The first part, tells us that the force resulting from the electric field $\vec{E}$ is directly proportional to $\vec{E}$:

$$\vec{F}_E = q\vec{E}, \tag{3.9}$$

where q is the proportionality constant, called electric charge, that encodes how strongly our object reacts to $\vec{E}$.

In words, Eq. 3.9 tells us that a charged object gets pushed in the direction in which the electric field points:[12]

[12] Take note that a positively charged object gets pushed in the direction of the electric field, while a negatively charged object gets pushed in the opposite direction. We will discuss this further in Section 4.1.5.

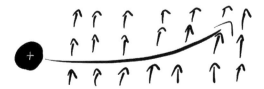

The second part of the Lorentz force law is a bit more complicated. While the force resulting from the magnetic field $\vec{B}$ is also directly proportional to $\vec{B}$, we need to take into account that only *moving* charged objects react to the presence of the magnetic field.[13]

[13] This can be understood using special relativity, which we will discuss in Chapter 6. From this perspective the statement here is a truism since "magnetic field" is simply how we call the relevant component of the electromagnetic field for an observer moving relative to the charge.

The quantity which we use to describe whether or not an object is moving is the velocity $\vec{v}$. In general, the velocity $\vec{v}$ is a vector and we therefore need to combine it somehow with the vector field $\vec{B}$ to yield a force $\vec{F}$. Since the force $\vec{F}$ is also a vector, we can't use the scalar product $\vec{v} \cdot \vec{B}$ which yields simply a number and not a vector.[14]

[14] As the name *scalar* product indicates, the result is a scalar, i.e. an ordinary number, not a vector. The scalar product is discussed in Appendix A.2.

However, we can use the cross-product $\vec{v} \times \vec{B}$ since it allows us to combine two vectors in such a way that they yield another vector.[15] This is indeed the correct idea and the second part of the Lorentz force law reads

[15] The cross product is the topic of Appendix A.3.

$$\vec{F}_B = q\vec{v} \times \vec{B}. \qquad (3.10)$$

In words, this means that while the force $\vec{F}_E$ resulting from the electric field points in the same direction as $\vec{E}$, the force resulting from the magnetic field $\vec{B}$ is perpendicular to $\vec{B}$ and the velocity $\vec{v}$.[16]

[16] As discussed in Appendix A.3, the definition of the cross-product is

$$\vec{a} \times \vec{b} = |\vec{a}||\vec{b}|\sin\theta \vec{n},$$

where $\vec{n}$ is the unit vector perpendicular to $\vec{a}$ and $\vec{b}$ and θ is the angle between $\vec{a}$ and $\vec{b}$. We can use this to *define* the magnetic field strength $|\vec{B}|$ through the force experienced by a test charge:

$$|\vec{B}| \equiv \frac{|\vec{F}_B|}{q|\vec{v}|\sin(\theta)}. \qquad (3.11)$$

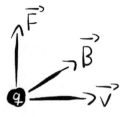

As a result the path of a charged object in a magnetic field is helical:[17]

[17] We will derive this explicitly in Section 4.2.2.

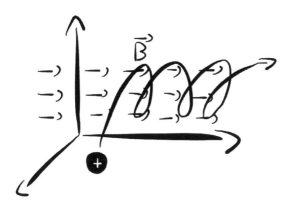

To be a bit more precise: we get such a helical path whenever the velocity $\vec{v}$ is not completely perpendicular to the magnetic field $\vec{B}$, since then there is a component of the velocity which moves the object further forward. In contrast, when the velocity $\vec{v}$ is completely perpendicular to the magnetic field $\vec{B}$, the trajectory is simply a circle:

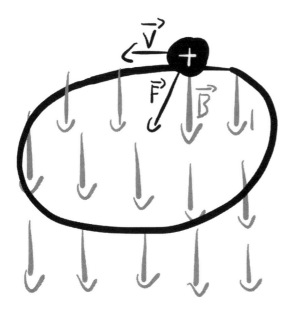

Maybe a second perspective helps to understand this better. In the following picture the X denote magnetic field vectors which point into the page you are looking at:

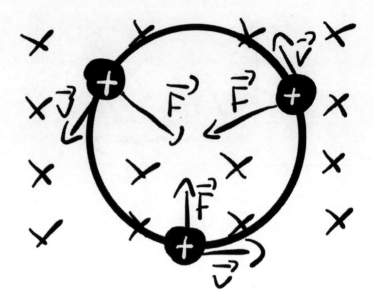

Let's summarize: The total force on a moving charged object is the sum of the force resulting from the electric field $\vec{E}$ and the magnetic field $\vec{B}$: $\vec{F}_{\text{res}} = \vec{F}_E + \vec{F}_B$. The general **Lorentz force law** therefore reads

$$\vec{F} = q(\vec{E} + \vec{v} \times \vec{B}). \tag{3.12}$$

Now, a natural question is: Where do nontrivial electric and magnetic field configurations come from? This is what Maxwell's equations tell us. In the following section, we will discuss Maxwell's equations in detail and start with the simplest one.

3.3 Gauss's law for the electric field

One of the most basic ideas in electrodynamics is that electric charges influence each other by modifying the surrounding electromagnetic field. Each electric charge generates a particular structure in the electromagnetic field and this, in turn, has a direct impact on other charges.[18]

Gauss's law for the electric field puts this statement into a mathematical form.[19] However, it only describes how charges influence the electric part of the electromagnetic field. If our charges are also moving, they additionally have a direct impact on the magnetic part of the electromagnetic field. This is described by "Gauss's law for the magnetic field". We will discuss this second Gauss's law in Section 3.4.[20]

In words, Gauss's law for the electric field describes how the electric field reacts to the presence of electric charges. How can we write this in mathematical terms using the fundamental notions introduced in Chapter 2?

Electric charge is described by the charge density ρ and the total charge inside some volume V is given by $\int_V \rho dV$. The basic statement described above means that if this total charge is non-zero, there will be a non-zero electric field strength $\vec{E}$. However, an equation like

$$a\vec{E} = \int_V \rho dV,$$

where a denotes a constant, makes no sense since $\int_V \rho dV$ on the right-hand side is simply a number while $\vec{E}$ is a vector. If we have a simple number on one side of an equation, we also need a number on the other side of the equation. Otherwise we are comparing apples with oranges.

Moreover, it seems reasonable that since the choice of the volume V plays a role on the right-hand side, it must play a role on the left-hand side too.

[18] The impact of a non-zero electric field strength on a given charge is described by the Lorentz force law (Eq. 1.5), which was the topic of the previous section.

[19] Take note that Gauss's Law $\neq$ Gauss's Theorem. Moreover, two of the four Maxwell equations (Eq. 1.4) are called Gauss's law. One describes the relationship between charges and the electric field and the second one describes the relationship between flowing charges and the magnetic field.

[20] We can already get a first glance here that it can lead to problems when we separate the electric and magnetic field. Namely, a charge which appears at rest for one observer, looks like a moving charge for a second observer who moves with some constant velocity relative to the charge. Hence, the first observer will calculate a non-zero electric field strength and a vanishing magnetic field strength, while the second observer calculates that the electric field vanishes but the magnetic field is non-zero. Thought experiments like this are what we usually consider in the context of special relativity and we will talk a bit more about this in Chapter 6.

We were in a similar situation when we wrote down the continuity equation in Section 3.1. There we wanted to write down an equation which connects the flow of electric charges (described by $\vec{J}$) with changes in the total amount of electric charge within some volume $\Delta t \frac{\partial}{\partial t} \int_V \rho dV$.

The idea which allowed us to connect the *vector* $\vec{J}$ with the *number* $\Delta t \frac{\partial}{\partial t} \int_V \rho dV$ was to use the total flux of electric charges through the surface of V (Eq. 2.14):

$$\text{flux of electric charge} \equiv \int_S \vec{J} \cdot d\vec{S}.$$

This flux represents the amount of electric charge which flows through the surface *per unit time*. Hence, if we multiply it by some interval Δt, we get the total amount of charge which passes through the surface during this interval. This amount is a simple number and depends directly on the choice of the volume V since it is an integral over the surface of the volume.

Writing this on one side of an equation and $\Delta t \frac{\partial}{\partial t} \int_V \rho dV$ on the other side makes sense. The equation we get this way is exactly the continuity equation (Eq. 3.6) discussed in Section 3.1.[21] The crucial idea is that we use exactly the same idea again. This means that we introduce the **flux of the electric field** through some surface S:

$$\text{flux of the electric field} \quad \phi \equiv \int_S \vec{E} \cdot d\vec{S}. \qquad (3.13)$$

This is simply a number and not a vector and therefore a good candidate for our left-hand side. Moreover, it also depends on the choice of the volume since S represents its surface.[22]

With this in mind, let's go back to what we discussed at the beginning of this section.

Our goal is to find an equation which describes that a nonzero charge within some volume has a direct impact on the structure of the electric field. The charge within some volume V is given by $\int_V \rho dV$ and we just argued that a reasonable object to represent the electric field in such an equation is the flux of the electric field (Eq. 3.13).

---

[21] Reminder: Eq. 3.6 reads

$$-\frac{\partial}{\partial t} \int_V \rho dV = \oint_S \vec{J} \cdot d\vec{S}.$$

[22] The main idea behind a flux integral like this is discussed in detail in Appendix A.9.

Putting these puzzle pieces together yields[23]

$$\oint_S \vec{E} \cdot d\vec{S} = \frac{1}{\epsilon_0} \int_V \rho \, dV, \qquad (3.14)$$

where ϵ_0 is a constant known as **electric permittivity** which describes how strongly the electric field reacts to the presence of charges. This is the **integral form of Gauss's law for electric fields**. In words, it tells us:

> The flux of the electric field passing through a closed surface is directly proportional to the amount of electric charge contained inside the volume.

It is important to keep in mind that the word "flux" is simply a statement about the length of the vectors at the boundary of our volume, i.e. on the surface. This means that if there is a lot of charge inside the surface, we get big vectors. If there is only a little charge inside, we get tiny vectors:

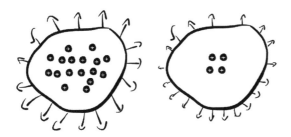

Moreover, the sign of the flux tells us in which direction the vectors are pointing. If the flux is positive, the vectors point outwards while if the flux is negative the vectors point inwards.

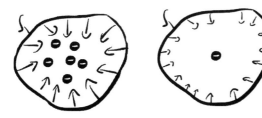

[23] Take note that we have $\oint$ since S is the boundary of the volume V and therefore a closed surface. Of course, the "derivation" here is by no means rigorous. As mentioned above, our goal is only to get a rough feeling for what electrodynamics is all about. Questions like where this equation and all other Maxwell equations really come from will be discussed later in detail. Historically they were, of course, simply deduced experimentally like, for example, the conservation of electric charge too. However, nowadays we can also understand the origin of Maxwell's equations and the conservation of electric charge from a more theoretical perspective. We will talk about this in Chapter 6.1.

Take note that there is really a vector at each point of the surface, but there is no way to draw them all. In addition, take note that if there is no charge inside the surface S or if there is an equal amount of positive and negative charges inside the volume, the net flux through the surface is zero.

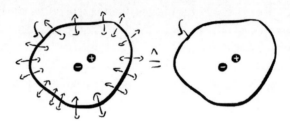

So, as promised above, Gauss's law simply puts the statement:

non-zero charge $\leftrightarrow$ non-zero electric field strength

into a mathematical form.[24]

Before we move on, three short comments on what we just learned.

[24] The explicit mathematical form of Gauss's law seems quite intimidating at first glance. However, it will become a lot clearer as soon as we use it below to calculate the electric field strength that results from a given charge distribution. We will do this in detail in Part II.

▷ An important difference between the flux of electric charge (the electric current per unit time) and the flux of the electric field is that the latter does not really describe the movement of anything. This can be quite confusing and has to do with the two ways we can understand vector fields like $\vec{J}$ and $\vec{E}$.[25] Both, $\vec{J}$ and $\vec{E}$ assign a vector to each point in space. However, for $\vec{J}$ these vectors represent the real movement of our electric charges. For the more abstract electric field $\vec{E}$, the arrows only represent how a charge *would* move if it were there.[26]

[25] We discussed this already near the end of Section 2.4.

[26] As already mentioned above, historically, physicists really thought that the arrows the electric field assigns to each point represent the movement of the "electric field substance". However, this idea has been experimentally falsified since such a real substance would lead to specific effects which were never observed.

▷ As mentioned above, there are two ways of thinking about Gauss's law (and also all other Maxwell equations). Either, we say that it describes how *the* electromagnetic field gets modified when there are electric charges present. In this view there is only one electromagnetic field which expands over all space but possibly has zero magnitude in some regions. Alternatively, we can say that it tells us what the electromagnetic fields produced by charges look like. Formulated differently, in this second view we imagine that each charge produces its own electromagnetic field. The second view is often more useful for concrete applications, but it is always important to keep in mind that there is really one electromagnetic field. The individual fields "produced" by charges are only a helpful way to think about the total field in simpler terms. The total field is what we can really observe in nature and it is always given by the sum over all these individual fields.

Now, before we discuss the second Maxwell equation we will rewrite Gauss's law in a more compact form. The steps are completely analogous to what we already discussed in Section 3.1 for the continuity equation. The result will be the differential form of Gauss's law.

We start with Gauss's law in integral form (Eq. 3.14) and then use again Gauss's theorem[27] to get a volume integral on both sides:

[27] Gauss's theorem is explained in Appendix A.13.

$$\int_S \vec{E} \cdot d\vec{S} = \frac{1}{\epsilon_0} \int_V \rho dV \quad \text{this is Eq. 3.14}$$

$$\downarrow \text{Gauss's theorem}$$

$$\therefore \int_V \nabla \cdot \vec{E} dV = \frac{1}{\epsilon_0} \int_V \rho dV$$

$$\downarrow$$

$$\therefore \int_V \left(\nabla \cdot \vec{E} - \frac{1}{\epsilon_0} \rho \right) dV = 0. \tag{3.15}$$

We can therefore conclude[28]

[28] Take note that this is the form of the first Maxwell equation, as given in Eq. 1.4.

$$\boxed{\nabla \cdot \vec{E} = \frac{\rho}{\epsilon_0}.} \tag{3.16}$$

This is the **differential form of Gauss's law for the electric field**.

What does this differential form tell us in plain language?

Since there are no integrals left in the final equation, the interpretation is a different one than for the integral form.[29] On the left-hand side we have the divergence of the electric field $\nabla \cdot \vec{E}$. In general, the divergence of a vector field gives us information about the tendency of the field to flow towards or away from a specific point.[30] On the right-hand side, we have the charge density ρ and the proportionality constant ϵ_0.[31]

[29] Reminder: the integral form of Gauss's law tells us that the flux of the electric field through a surface is directly proportional to the electric charge contained in the volume.

[30] The divergence of a vector field is discussed in detail in Eq. A.11.

[31] Reminder: the proportionality constant ϵ_0 is known as the permittivity and encodes how strongly the electromagnetic field reacts to the presence of charges.

With this in mind, we can say that the differential form of Gauss's law for the electric field tells us:

> The structure of the electric field generated by electric charges is such that it converges upon negative charges and diverges from positive charges.

We can also recast this statement in pictorial form:

Formulated differently, a positive charge implies a positive divergence of the electric field. This, in turn, means that the electric field vectors point away from the location of the positive charge. A negative charge implies a negative divergence of the electric field. Hence, the electric field vectors point towards the location of the negative charge.

Another important aspect of the differential form of Gauss's law is that it tells us that whenever the divergence of the electric field is non-zero, there must necessarily be a non-zero electric charge. This means that as soon as we find the particular structure in the electric field associated with a non-zero divergence, we know immediately that there is non-zero electric charge.

With this in mind, we can say that the crucial difference between the integral and the differential forms of our equations is that the latter make statements about the field structure and charge distribution at individual points. In contrast, the integral form always makes statements about the structure of the electric field on complete surfaces and on the total charge within some volume.

However, take note that the interpretations of the two forms of Gauss's law are, of course, not really independent. For the integral form, we can imagine that our volume gets smaller and smaller until it only represents the neighborhood of a single point. We can then make our volume gradually larger and scan the complete neighborhood of a given charge. At each step, we can use the integral form of Gauss's law to calculate the flux of the electric field. For a positive charge we find a positive flux. For a negative charge, we find a negative flux. A positive flux tells us that the vectors of the electric field point away from the single point where we started. A negative flux tells us that the vectors point towards this point. In this sense, the integral form also contains the statement that the field structure converges upon negative charges and diverges from positive charges.

But then why do we write our equations in two different ways?

Well, both the integral form and the differential form have important advantages depending on the problem at hand. The integral form is especially useful for macroscopic problems which posses a high degree of symmetry, e.g. when there is a rotationally symmetric charge distribution.[32] The differential form is useful, for example, when we want to talk about electromagnetic waves.[33] Moreover, the differential form is often more useful when we want to evaluate Maxwell's equations numerically for a specific problem and also for fundamental considerations like in the context of special relativity.

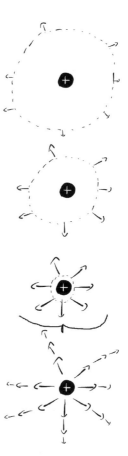

Figure 3.1: By gradually scanning the neighborhood of a given charge with different surfaces, we can deduce the electric field configuration. An important observation is that we always get the same total flux for all surfaces around a given charge distribution. This means that we have smaller arrows if we are farther away from the charges since the same amount of flux must be distributed along a larger surface.

[32] We will see this explicitly in Chapter 4.

[33] Electromagnetic waves are particular structures in the electromagnetic field which can travel without any charges nearby. We will talk about such waves in Chapter 5.

Now, let's move on to the second Maxwell equation which is really similar to the first one.

3.4 Gauss's law for the magnetic field

If Gauss's law for the magnetic field were completely analogous to Gauss's electric law discussed in the previous section, it would read[34]

$$\oint_S \vec{B} \cdot \vec{dS} = \frac{1}{\tilde{\epsilon}_0} \int_V \tilde{\rho} \, dV, \quad (3.17)$$

where $\tilde{\epsilon}_0$ is some new proportionality constant and $\tilde{\rho}$ the magnetic charge density. However, so far no non-zero magnetic charge density has ever been observed.[35] Since no magnetic charge has ever been observed, for all practical purposes we have $\tilde{\rho} = 0$ and the **integral form of Gauss's law for the magnetic field** reads

$$\boxed{\oint_S \vec{B} \cdot \vec{dS} = 0,} \quad (3.18)$$

Completely analogous to what we discussed for Gauss's law for the electric field in the previous section, we have on the left-hand side the flux of the magnetic field through the closed surface S. So in words, it tells us

> The flux of the magnetic field passing through any closed surface is zero.

The statement "no magnetic charge has ever been observed" can be quite confusing since where then does a non-zero magnetic field strength come from?

The short answer is that a non-zero magnetic field strength can be generated by *moving* electric charges and also by a changing electric field. We will discuss this in more detail in the following sections.

If we could produce a non-zero magnetic charge density, we would have to replace Eq. 3.18 with Eq. 3.17, which in words

[34] Reminder: Gauss's law for the electric field (Eq. 3.14) reads

$$\oint_S \vec{E} \cdot \vec{dS} = \frac{1}{\epsilon_0} \int_V \rho \, dV.$$

[35] Formulated differently, no magnetic monopole has ever been observed. Magnetic monopoles would act as a source for magnetic field, analogous to how electric charges generate non-zero electric field strengths. Even today, quite a few physicist believe in the existence of magnetic monopoles since they appear inevitably in so-called Grand Unified Theories and, in addition, would help to explain why electric charge is quantized. However, the modern point of view is that magnetic monopoles, if they exist, are not point charges like electric charges, but rather special extended configurations of the electromagnetic field. These special configurations are stable thanks to their special (topological) structure and happen to act from a distance exactly like a magnetic monopole.

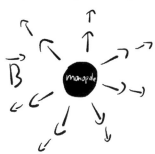

tells us: The flux of the magnetic field passing through a closed surface is directly proportional to the amount of magnetic charge contained inside the surface. Technically, this statement is true in general, since it also holds for a zero magnetic charge, which is exactly the statement from above.

Formulated differently, if a non-zero magnetic charge density would exist, there would be a non-zero flux of the magnetic field through a closed surface. However, since no non-zero magnetic charge density has ever been observed, the flux through a closed surface happens to be always zero.

Next, completely analogous to what we did in the previous two sections, we can derive the differential form of Gauss's law for the magnetic field.

We start again with Gauss's law in integral form (Eq. 3.18) and then use Gauss's theorem[36] to transform the surface integral into a volume integral:[37]

$$\oint_S \vec{B} \cdot d\vec{S} = 0 \qquad \text{this is Eq. 3.18}$$

$$\text{Gauss's theorem}$$

$$\therefore \int_V \nabla \cdot \vec{B} \, dV = 0. \tag{3.19}$$

We can therefore conclude[38]

$$\boxed{\nabla \cdot \vec{B} = 0.} \tag{3.20}$$

This is the **differential form of Gauss's law for magnetic field**.

In words, this form of Gauss's law tells us

> The divergence of the magnetic field is always zero.

Therefore, once more the differential form makes an explicit statement about configurations at specific locations while the integral form makes a statement about configurations on surfaces.

[36] As already mentioned several times above, Gauss's theorem is explained in Appendix A.13.

[37] Take note that this trick is necessary since we cannot make a statement about the integrand without it. Especially take note that we cannot conclude $\vec{B} = 0$ since we have in the integral form the *projection* of the magnetic field onto the surface as the integrand. This is what the scalar product $\vec{B} \cdot d\vec{S}$ encodes.

[38] Take note that this is the form of the third Maxwell equation, as given in Eq. 1.4.

We have on the left-hand side the divergence of the magnetic field $\nabla \cdot \vec{B}$.[39] The divergence of a vector field describes how the field flows towards or away from a given point.[40]

Since the right-hand side of our equation is zero, we can conclude that the tendency of the magnetic field to flow *towards* a point P is always exactly equal to its tendency to flow *away* from P. Any "inflow" of the magnetic field is always accompanied by an "outflow" of exactly the same magnitude.[41]

[39] Reminder: the differential form of Gauss's law for electric field reads
$$\nabla \cdot \vec{E} = \frac{\rho}{\epsilon_0}.$$

[40] This is discussed in detail in Appendix A.11.

[41] Once more it is important to keep in mind that nothing is really flowing when we speak about the flow or flux of the electric and magnetic field. This language is only convenient to get some intuitive feeling for how the vectors that make up the vector field are arranged.

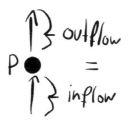

Here's an example of a non-trivial field configuration where this is indeed the case:[42]

[42] Of course, the statement "inflow"="outflow" is trivially true when the vectors have zero length, i.e. for a vanishing magnetic field strengths.

A net non-zero flow in either direction would only be possible with an isolated magnetic charge, analogous to how the divergence of an electric field can be zero if there is an electric charge present.

Formulated differently, since there are no sources or sinks for the magnetic field (i.e. magnetic monopoles) at each point the amount of magnetic field flux entering a point must be exactly equal to the amount of magnetic field flux leaving the point.[43]

Once more, before we move on, a few comments on what we just learned:

▷ You probably wonder why people talk about magnets but above, we argued that there is no magnetic charge. We can, of course, produce a non-zero magnetic field strength in systems. Objects which generate a non-zero magnetic field strength are what we call magnets. However, the structure of the magnetic field generated this way is quite different from the structure that a real fundamental magnetic charge would produce. We will see in Section 3.6 that we can generate a non-zero magnetic field strength using a varying electric field or by using moving electric charges. The crucial difference between the structure of the field generated by a fundamental charge and the structure generated by another field or moving electric charges is visible in the divergence of the vector field. If the divergence is non-zero, the structure was created by a fundamental charge. If the divergence is zero, the structure was created by a different field (or a current).[44]

The non-zero magnetic field strengths generated by conventional magnets are generated by the billions of individual electrons which exist inside any macroscopic object. Each electron orbits around a nucleus and possesses a non-zero internal angular momentum.[45] As a result, we have billions of tiny currents inside any macroscopic object and each of these currents generates a tiny non-zero magnetic field strength.

But this happens in any object and for any material. The special thing about magnets is that in these objects the majority of electrons rotate around approximately the same axis. This way the hundreds of tiny magnetic field strengths add up to a measurable macroscopic magnetic field strength.

[43] As we discuss in Appendix A.11, the divergence is a scalar field which assigns a number to each point in space. A non-zero divergence always indicates that there is a source or sink for the corresponding vector field. A vector field with divergence zero is known as a **solenoidal vector field**.

[44] Roughly think:

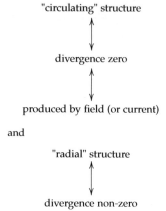

All this is described by the Ampere-Maxwell law, which we will discuss in Section 3.6.

[45] This internal angular momentum is known as **spin**. Naively, you can imagine that each electron rotates like a little spinning top and there is no way to make them stop. The spin of each elementary particle is a defining and unchangeable feature of it, analogous to the labels: electric charge and mass.

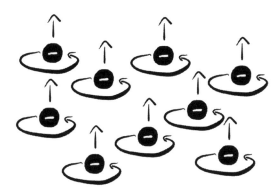

In contrast, in a normal object, the electrons circulate around random axes and the tiny magnetic field strengths average out to zero.[46]

[46] Take note that usually there is no non-zero macroscopic electric field strength due to the tiny electric field strengths of the electrons since they are canceled through the electric field strengths of the protons.

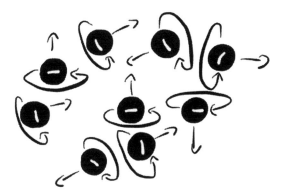

▷ Take note that we can also achieve a non-zero electric field with zero divergence. This is possible since a non-zero electric field strength can not only be generated by electric charges but also by a changing magnetic field. This is described by Faraday's law, which is the topic of the next section. The structure of the electric field generated this way is characterized by a vanishing divergence. This means that the structure of the electric field generated by a changing magnetic field circulates back on itself, while the structure generated by an electric charge has a non-vanishing divergence and does not circulate back on itself.

With this in mind, let's move on to the third Maxwell equation which describes the interplay between the electric and the magnetic field.

3.5 Faraday's law

As already mentioned in the previous section, non-zero field strengths can not only be produced by charges but also by other fields. For example, a non-zero electric field strength can be generated when there is a changing magnetic field.

How can we put the idea

$$\text{changing magnetic field} \rightarrow \text{nontrivial electric field}$$

into a mathematical form?

We want an equation with the changing magnetic field on one side and the resulting electric field configuration on the other side.

Therefore, first of all we need to understand how we can encode that a magnetic field is changing. So far, when our fields appeared in the integral form of an equation, they appeared in the form of a surface integral (or volume integral after we used Gauss's theorem). For example, in Gauss's law for the magnetic field we have $\oint_S \vec{B} \cdot d\vec{S}$ on the left-hand side.[47]

[47] Recall that such a surface integral describes the flux of the corresponding vector field through the surface S.

Hence, a good first guess is to describe our magnetic field using a surface integral again. However, there are two things that we need to take care of.

▷ We want a description in which a non-zero electric field strength shows up when the magnetic field is changing. The correct mathematical notion for this task are time-derivatives. If the time-derivative of the magnetic field is non-zero, this means that the magnetic field is changing over time. Hence, our first puzzle piece is $\oint_S \frac{\partial}{\partial t} \vec{B} \cdot d\vec{S}$.

FUNDAMENTAL EQUATIONS 75

▷ In the previous two sections our surface was always a closed surface since it represented the surface of a specific volume. However, the flux of the magnetic field through a closed surface is always zero and hence cannot be used here. This is what Gauss's law for the magnetic field tells us.[48] But we can try to write down an equation for the time-derivative of the magnetic flux through an *arbitrary* (not necessarily closed) surface S:

$$\ldots = \int_S \frac{\partial}{\partial t} \vec{B} \cdot d\vec{S}. \qquad (3.21)$$

There is no reason why this should vanish in general.

[48] Just imagine that we put the time-derivative in front of the integral. Then we simply get the time-derivative of the flux. But the flux of the magnetic field is zero for any closed surface.

The second puzzle piece that we need is something which describes the structure of the electric field that results from such a changing magnetic field.

A first hint is that the left-hand side must depend on our choice of the surface S too. However, a volume integral over $\vec{E}$ is not a good idea since then S would represent the surface of the volume. As a result S would be a closed surface and, as mentioned above, Gauss's law for the magnetic field tells us that the integral over the magnetic field for any closed surface vanishes.

But maybe this time we can use the boundary of the surface S instead?

The boundary of a surface is simply its contour:[49]

[49] It is conventional to denote the boundary of a given geometrical object G by δG. For example, the boundary of a volume V is δV and the boundary of some surface S is δS.

With this in mind, we try the contour integral over the electric field $\oint_C \vec{E} \cdot d\vec{l}$ on the left-hand side, where C denotes the contour of the surface S:

$$\oint_C \vec{E} \cdot \vec{dl} = \int_S \frac{\partial}{\partial t} \vec{B} \cdot \vec{dS}. \qquad (3.22)$$

This is *almost* correct. Only a minus sign is missing. Correctly, Faraday's law in integral form reads

$$\boxed{\oint_C \vec{E} \cdot \vec{dl} = -\int_S \frac{\partial}{\partial t} \vec{B} \cdot \vec{dS}.} \qquad (3.23)$$

The contour integral on the left-hand side describes the circulation of the electric field.[50] Therefore, the relative sign between the right-hand and left-hand side encodes in which direction the resulting electric field circulates.[51] The minus sign is an empirical result known as **Lenz's law**.

In some sense, Lenz's law is not very surprising since it simply encodes the usual wisdom that Nature is lazy.[52] We will see in the next section that an electric current generates a magnetic field. Thus, if there are electrons at the location of the contour they will start circulating and this, in turn, will generate a specific structure in the magnetic field. The direction in which they circulate determines if they make it easier for the magnetic field to change or if they make it harder. The relative minus sign tells us that they make it harder, i.e. they generate a magnetic field structure which tries to stop the original change in the magnetic field. So when the magnetic flux through the surface is increasing, the electrons circulate in such a way that they generate an additional magnetic flux in the opposite direction of the original flux.

In other words, nature tries to make sure that everything stays as it is and actively works against changes e.g., in the magnetic field.[53]

To summarize, Faraday's law tells us

> A changing magnetic field flux generates a circulating structure in the electric field.

---

[50] The circulation of a vector field is discussed in more detail in Appendix A.7 and also below.

[51] Again, nothing is really circulating. This is only a convenient way to talk about the structure of the vectors of the electric field. However, if there really are electrons present in the system, they will indeed start circulating.

[52] "Nature is lazy" is often a helpful guiding principles in physics. For example, it allows us to understand why the Lagrangian formalism works. See, for example,

Jennifer Coopersmith. *The lazy universe : an introduction to the principle of least action.* Oxford University Press, Oxford New York, NY, 2017. ISBN 9780198743040

[53] However, take note that the change in the magnetic field is only made harder if there are really electrons. If no electrons are present, Lenz's law only tells us that the direction in which the electric field circulates is such that the corresponding change in the magnetic flux *would* get harder if there were electrons.

In the previous sections we always dealt with surface integrals which could be interpreted as the flux of our fields. Now, what's the physical meaning of a contour integral like $\oint_C \vec{E} \cdot d\vec{l}$?[54]

[54] As mentioned above, in somewhat abstract terms such a contour integral describes the circulation of the field. This point of view is discussed in Appendix A.7.

We can understand this circulation of the electric field by recalling how the electric field and the resulting force on a test charge are related. The connection between the two is given by the Lorentz force law (Eq. 3.12)

$$\vec{F} = q\vec{E}. \qquad (3.24)$$

Using this, we can calculate

$$\oint_C \vec{E} \cdot d\vec{l} = \oint_C \frac{\vec{F}}{q} \cdot d\vec{l} \qquad \text{this is Eq. 3.24 integrated}$$

$$\text{↷ the charge is a constant}$$

$$= \frac{1}{q} \oint_C \vec{F} \cdot d\vec{l}$$

$$\text{↷}$$

$$\equiv \frac{W}{q}, \qquad (3.25)$$

where W denotes the work done by the electric field if we move a test charge q along the path C.

Therefore, we can conclude that the circulation of the electric field represents the energy for each unit of charge moving along the contour C.[55]

[55] An analogous statement is true for the circulation of the magnetic field which will be important in the next section.

Next, analogous to what we did in the previous sections, we will derive the differential form of Faraday's law.

This time, the crucial trick is not Gauss's theorem but a closely related trick known as Stokes' theorem.

In the previous sections, we always had a surface integral which we wanted to turn into a volume integral and this is exactly what Gauss's theorem allows us to do. Now, we have on the left-hand side a contour integral and on the right-hand side a surface integral. Therefore, we need a trick that allows us to convert a contour integral into a surface integral and this is exactly what Stokes' theorem allows us to do.[56] Stokes' theorem tells us[57]

$$\oint_C \vec{E} \cdot \vec{dl} = \int_S \nabla \times \vec{E} \cdot \vec{dS} \qquad (3.26)$$

for any vector field $\vec{E}$ and therefore also for our electric field. S denotes a surface for which the contour C is a boundary.

[56] As a reminder: we want the same kind of integral on both sides of the equation since this allows us to write the equation without any integral.

[57] Stokes' theorem is discussed in Appendix A.15.

Using this theorem, we can derive the differential form of Faraday's law:

$$\oint_C \vec{E} \cdot \vec{dl} = -\int_S \frac{\partial}{\partial t} \vec{B} \cdot \vec{dS} \qquad \text{this is Faraday's law, Eq. 3.23}$$

$$\downarrow \text{Stokes' theorem}$$

$$\therefore \int_S \nabla \times \vec{E} \cdot \vec{dS} = -\int_S \frac{\partial}{\partial t} \vec{B} \cdot \vec{dS}.$$

Since this equation holds for any surface S, we can conclude[58]

[58] Take note that this is the form of the fourth Maxwell equation, as given in Eq. 1.4.

$$\boxed{\nabla \times \vec{E} = -\frac{\partial}{\partial t} \vec{B}.} \qquad (3.27)$$

This is the **differential form of Faraday's law**.

The object $\nabla \times \vec{E}$ appearing on the left-hand side is the curl of the electric field. In general, a curl always tells us about the tendency of a vector field to circulate around a given point.[59]

[59] The curl of vector fields is discussed in general in Appendix A.12.

Therefore, the differential form of Faraday's law tells us in words[60]

[60] In contrast, the two Maxwell equations we discussed previously involved the divergence of the fields and describe how radial structures emerge in them.

> A changing magnetic field generates a non-zero curl of the electric field.

Next, we are finally ready to discuss the final Maxwell equation.

3.6 The Ampere-Maxwell law

The Ampere-Maxwell law is quite similar to Faraday's law but the roles of the electric and magnetic fields are switched. This means that the Ampere-Maxwell law describes how a changing electric field generates a non-zero magnetic field strength.[61] If it were completely analogous to Faraday's law it would read

$$\oint_C \vec{B} \cdot d\vec{l} = -\text{const.} \times \int_S \frac{\partial}{\partial t} \vec{E} \cdot d\vec{S}. \qquad (3.28)$$

While this equation is correct for some systems, there is an additional effect that we need to take into account.

In the previous sections, we have learned that a non-zero electric field strength can be generated through a non-zero charge density or a changing magnetic field. Now, Eq. 3.28 tells us that a non-zero magnetic field strength can be generated through a changing electric field. Is there any other way to generate a non-zero magnetic field strength?

It turns out that we can also generate a non-zero magnetic field strength using *moving* electric charges:[62]

[61] Reminder: Faraday's law describes how a changing magnetic field generates a non-zero electric field strength and reads (Eq. 3.23)

$$\oint_C \vec{E} \cdot d\vec{l} = -\int_S \frac{\partial}{\partial t} \vec{B} \cdot d\vec{S}.$$

[62] In fact, the electric current I and the current density $\vec{J}$ are the only ingredients we haven't used so far in Maxwell's equations.

Adding the current to Eq. 3.28 yields

$$\boxed{\oint_C \vec{B} \cdot d\vec{l} = \mu_0 \left(I_{\text{enc}} + \epsilon_0 \int_S \frac{\partial}{\partial t} \vec{E} \cdot d\vec{S} \right).} \qquad (3.29)$$

This is the **integral form of the Ampere-Maxwell law**. μ_0 and ϵ_0 are proportionality constants known as **vacuum permeability**

[63] Take note that $\mu_0 \epsilon_0 = \frac{1}{c^2}$, where c denotes the speed of light. We will discuss this in detail in Appendix 6.

and **vacuum permittivity**.[63] Moreover, I_{enc} denotes the electric current *enclosed* by the contour C.

On the left-hand side we have a contour integral which describes the circulation of the field $\vec{B}$ along the contour. On the right-hand side, we have the current enclosed by the contour and the flux of the electric field through a surface S of which the contour C is the boundary.

Therefore, in words the Ampere-Maxwell law tells us[64]

[64] Of course, there are systems where we have both a changing electric field and a non-zero current. In this case, the effects of the two either cancel or amplify each other.

> A changing electric field flux or a non-zero electric current generates a circulating structure in the magnetic field.

So while the Ampere-Maxwell law may seem intimidating at first glance, the basic message is quite simple. Most importantly, we can see that its structure is completely analogous to the structure of the other Maxwell equations.

Finally, analogous to what we did in the previous sections, we can derive the differential form of the Ampere-Maxwell law.

The crucial trick is exactly the same as in the previous section (Stokes' theorem) since again, we want to transform a contour integral into a surface integral:

$$\oint_C \vec{B} \cdot \vec{dl} = \int_S \nabla \times \vec{B} \cdot \vec{dS}. \tag{3.30}$$

With this trick in mind, we can derive the differential form of the Ampere-Maxwell law immediately:

$$\oint_C \vec{B} \cdot d\vec{l} = \mu_0 \left(I_{\text{enc}} + \epsilon_0 \int_S \frac{\partial}{\partial t} \vec{E} \cdot d\vec{S} \right)$$

this is Eq. 3.29

↷ Stokes' theorem

$$\therefore \int_S \vec{\nabla} \times \vec{B} \cdot d\vec{S} = \mu_0 \left(I_{\text{enc}} + \epsilon_0 \int_S \frac{\partial}{\partial t} \vec{E} \cdot d\vec{S} \right)$$

↷ def. of $\vec{J}$ in Eq. 2.14

$$\therefore \int_S \vec{\nabla} \times \vec{B} \cdot d\vec{S} = \mu_0 \left(\int_S \vec{J} \cdot d\vec{S} + \epsilon_0 \int_S \frac{\partial}{\partial t} \vec{E} \cdot d\vec{S} \right)$$

↷

$$\therefore \int_S \vec{\nabla} \times \vec{B} \cdot d\vec{S} = \mu_0 \int_S \left(\vec{J} + \epsilon_0 \frac{\partial}{\partial t} \vec{E} \right) \cdot d\vec{S}. \quad (3.31)$$

Again, this equation holds for any surface S and we can therefore conclude

$$\boxed{\vec{\nabla} \times \vec{B} = \mu_0 \left(\vec{J} + \epsilon_0 \frac{\partial}{\partial t} \vec{E} \right).} \quad (3.32)$$

This is the **differential form of the Ampere-Maxwell law**.

In words, it tells us[65]

> A changing electric field or a non-zero current density generates a non-zero curl in the magnetic field.

[65] Again, of course, there are systems where we have both a changing electric field and a non-zero current density. In this case, the effects of the two either cancel or amplify each other.

Before we move on and discuss applications of Maxwell's equations, there are two additional equations we should talk about: the wave equations of electrodynamics. In some sense, they are not as fundamental as Maxwell's equations since we can derive them by starting with Maxwell's equations. However, they reveal an important feature of electrodynamics which is otherwise somewhat hidden.[66]

The feature I'm talking about is that there can be **electromagnetic waves**. We have seen in the previous sections that a changing magnetic field can generate a non-zero electric field strength and vice versa. The structure which is generated in the electric field this way is itself changing over time and therefore influences the magnetic field:

[66] The same is true for the continuity equation. We can derive it using Maxwell's equations but it is still an important equation because it makes the conservation of electric charge completely transparent.

changing $\vec{B}$ ⟶ changing $\vec{E}$ ⟶ changing $\vec{B}$ ⟶ ...

This way, nontrivial electric and magnetic field configurations can travel long distances and this is what we call an electromagnetic wave. No messenger, no medium, no electric charges are needed. An electromagnetic wave can propagate completely on its own, even through a vacuum. The most famous example of this phenomena is sunlight which travels from the sun to the earth.

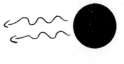

The equations which describe electromagnetic waves are the topic of our final section in this chapter.

3.7 The wave equations

So far, we have no equation that tells us how our electric and magnetic fields evolve as time passes. In other words, we have no equation of motion for our fields $\vec{E}$ and $\vec{B}$. Such an equation of motion has a time derivative on one side and other terms, possibly with spatial derivatives, containing the same field on the other side. In this section, we will combine Maxwell's equation to derive exactly such equations.[67]

[67] As already mentioned above, for $\vec{E}$ and $\vec{B}$ these equations of motion are known as wave equations. This name comes about since solutions of these equations look and behave like waves. We will talk more about this in Chapter 5.

To derive the equation of motion for the electric field $\vec{E}$, we start by taking the curl on both sides of Faraday's law[68]

$$\nabla \times \vec{E} = -\frac{\partial}{\partial t}\vec{B} \qquad \text{Faraday's law, Eq. 3.27}$$

↝ taking the curl on both sides

$$\therefore \quad \nabla \times \nabla \times \vec{E} = -\nabla \times \frac{\partial}{\partial t}\vec{B}$$

↝ partial deriviatives commute

$$\therefore \quad \nabla \times \nabla \times \vec{E} = -\frac{\partial}{\partial t}\nabla \times \vec{B}. \qquad (3.33)$$

[68] In the last step we use that partial derivatives are symmetric:

$$\frac{\partial}{\partial x}\frac{\partial}{\partial t} = \frac{\partial}{\partial t}\frac{\partial}{\partial x}.$$

This is known as Schwarz's theorem and is one of the basic results in calculus.

Next, we can rewrite the left-hand side using the vector identity[69]

$$\nabla \times \nabla \times \vec{E} = \nabla(\nabla \cdot \vec{E}) - \nabla^2 \vec{E} \qquad (3.35)$$

which holds for any vector and therefore also for our electric vector field $\vec{E}$. In addition, on the right-hand side we can use the Ampere-Maxwell law[70] to get an equation which only depends on the electric field:

$$\nabla \times \nabla \times \vec{E} = -\frac{\partial}{\partial t}\nabla \times \vec{B} \qquad \text{this is Eq. 3.33}$$

$$\circlearrowright \text{ Eq. 3.35}$$

$$\nabla(\nabla \cdot \vec{E}) - \nabla^2 \vec{E} = -\frac{\partial}{\partial t}\nabla \times \vec{B}$$

$$\circlearrowright \text{ Eq. 3.32}$$

$$\nabla(\nabla \cdot \vec{E}) - \nabla^2 \vec{E} = -\frac{\partial}{\partial t}\left(\mu_0\left(\vec{J} + \epsilon_0\frac{\partial}{\partial t}\vec{E}\right)\right). \qquad (3.36)$$

We can simplify this further using Gauss's law for the electric field[71] and by restricting ourselves to systems with no charge density and no current present ($\rho = 0$ and $\vec{J} = 0$)

$$\nabla(\nabla \cdot \vec{E}) - \nabla^2 \vec{E} = -\frac{\partial}{\partial t}\left(\mu_0\left(\vec{J} + \epsilon_0\frac{\partial}{\partial t}\vec{E}\right)\right) \qquad \text{Eq. 3.36}$$

$$\circlearrowright \text{ Eq. 3.16}$$

$$\nabla(\frac{\rho}{\epsilon_0}) - \nabla^2 \vec{E} = -\frac{\partial}{\partial t}\left(\mu_0\left(\vec{J} + \epsilon_0\frac{\partial}{\partial t}\vec{E}\right)\right)$$

$$\circlearrowright \rho = 0, \vec{J} = 0$$

$$-\nabla^2 \vec{E} = -\mu_0 \epsilon_0 \frac{\partial^2}{\partial t^2} \vec{E}. \qquad (3.37)$$

Therefore, we can conclude

$$\boxed{\nabla^2 \vec{E} = \mu_0 \epsilon_0 \frac{\partial^2}{\partial t^2} \vec{E}.} \qquad (3.38)$$

This is the **wave equation for the electric field**.

Starting with the Ampere-Maxwell law and then following completely analogous steps yields

$$\boxed{\nabla^2 \vec{B} = \mu_0 \epsilon_0 \frac{\partial^2}{\partial t^2} \vec{B}.} \qquad (3.39)$$

This is the **wave equation for the magnetic field**.

[69] This is discussed in Appendix A.16. Here ∇^2 is known as the Laplacian operator

$$\nabla^2 \vec{E} = \begin{pmatrix} \left(\frac{\partial^2}{\partial x^2} + \frac{\partial^2}{\partial y^2} + \frac{\partial^2}{\partial z^2}\right) E_x \\ \left(\frac{\partial^2}{\partial x^2} + \frac{\partial^2}{\partial y^2} + \frac{\partial^2}{\partial z^2}\right) E_y \\ \left(\frac{\partial^2}{\partial x^2} + \frac{\partial^2}{\partial y^2} + \frac{\partial^2}{\partial z^2}\right) E_z \end{pmatrix}.$$

In contrast

$$\nabla(\nabla \cdot \vec{E}) = \nabla\left(\frac{\partial E_x}{\partial x} + \frac{\partial E_y}{\partial y} + \frac{\partial E}{\partial z}\right)$$

$$= \begin{pmatrix} \partial_x\left(\frac{\partial E_x}{\partial x} + \frac{\partial E_y}{\partial y} + \frac{\partial E}{\partial z}\right) \\ \partial_y\left(\frac{\partial E_x}{\partial x} + \frac{\partial E_y}{\partial y} + \frac{\partial E}{\partial z}\right) \\ \partial_z\left(\frac{\partial E_x}{\partial x} + \frac{\partial E_y}{\partial y} + \frac{\partial E}{\partial z}\right) \end{pmatrix}$$

$$\neq \nabla^2 \vec{E}. \qquad (3.34)$$

[70] Reminder: the Ampere-Maxwell law reads

$$\nabla \times \vec{B} = \mu_0\left(\vec{J} + \epsilon_0 \frac{\partial}{\partial t}\vec{E}\right)$$

[71] Reminder: Gauss's law for the electric field reads

$$\nabla \cdot \vec{E} = \frac{\rho}{\epsilon_0}.$$

Before we wrap up and summarize what we have learned in this chapter, two short comments on these wave equations.

▷ In Chapter 5, we will discuss solutions of these wave equations explicitly. The main result will be that these solutions really look and behave like waves, which is why we call these equations *wave* equations.

▷ At first it may seem strange that we get two separate equations for the electric and the magnetic field. This can be especially confusing if you recall that at the end of the last section I argued that an electromagnetic wave propagates through the interplay of the electric and magnetic field. However, the electric and magnetic fields always influence each other, even if we have here two seemingly independent equations since Maxwell's equations are always valid. Hence, a specific solution of the wave equation for the electric field *automatically* implies a corresponding solution of the wave equation for the magnetic field. In this sense, the two equations only appear to be independent. But in fact, they are directly connected via Maxwell's equations. We will discuss this explicitly in Section 5.2.

Now it's time to summarize what we have learned in this chapter.

3.8 Summary

The first equation we talked about is the **continuity equation** in integral form (Eq. 3.6) and differential form (Eq. 3.8)

$$-\frac{\partial}{\partial t}\int_V \rho dV = \int_S \vec{J}\cdot d\vec{S} \quad \underset{\text{theorem}}{\overset{\text{Gauss's}}{\Longrightarrow}} \quad \frac{\partial}{\partial t}\rho + \nabla\cdot\vec{J} = 0.$$

This equation tells us that whenever the amount of electric charge in some region changes, exactly the missing or additional amount of electric charge must have passed the boundary of the region. Formulated differently, when the amount of electric charge in some region gets larger over time, the continuity equation tells us that there must be a net inflow of electric charge. Conversely, when the amount of electric charge gets smaller, there must be a net outflow. This is necessarily the case because electric charge is conserved, i.e. cannot be produced or destroyed.

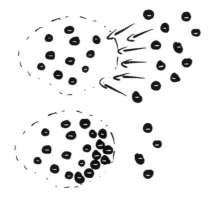

Next, we talked about the **Lorentz force law** (Eq. 3.12)

$$\vec{F} = q(\vec{E} + \vec{v} \times \vec{B}). \qquad (3.40)$$

This equation describes how electrically charged objects react to the presence of a non-zero electric or magnetic field strength.

It tells us that a positively charged object gets pushed in the direction of the electric field. A negatively charged object gets pushed in the opposite direction.

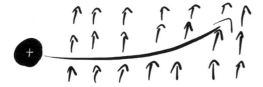

In addition, the Lorentz force law tells us that the magnetic field only influences *moving* charged objects. The force resulting from a non-zero magnetic field strength is perpendicular to both the direction of the magnetic field $\vec{B}$ and the direction in which the object moves $\vec{v}$.

Afterwards, we started talking about Maxwell's equations. They describe how the electric and magnetic fields react to the presence of charged objects and how they influence each other.

We started with **Gauss's law for the electric field** and talked about its integral form (Eq. 3.14) and its differential form (Eq. 3.16)

$$\oint_S \vec{E} \cdot d\vec{S} = \frac{1}{\epsilon_0} \int_V \rho dV \quad \overset{\text{Gauss's}}{\underset{\text{theorem}}{\Longrightarrow}} \quad \nabla \cdot \vec{E} = \frac{\rho}{\epsilon_0}.$$

This equation tells us that the electric field strength around some region is directly proportional to the amount of charge contained within the region.

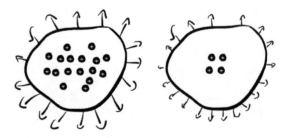

Moreover, it tells us that the electric field points radially away from any positively charged object and radially towards any negatively charged object.[72]

---

[72] This is non-trivial since, in principle, it could also flow, for example, circularly *around* any given charge.

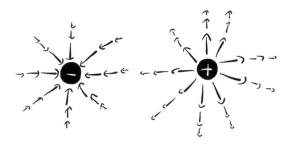

The second Maxwell equation we talked about was Gauss's law for the magnetic field (Eq. 3.18, Eq. 3.20)

$$\oint_S \vec{B} \cdot d\vec{S} = 0 \quad \underset{\text{theorem}}{\overset{\text{Gauss's}}{\Longrightarrow}} \quad \nabla \cdot \vec{B} = 0.$$

In words, it tells us that the magnetic field never points radially towards or away from a single point since there are no magnetic monopoles.

Then we talked about the two most complicated Maxwell equations which both describe how the electric and magnetic fields influence each other.

In particular, **Faraday's law** (Eq. 3.23, Eq. 3.27)

$$\oint_C \vec{E} \cdot d\vec{l} = -\int_S \frac{\partial}{\partial t} \vec{B} \cdot d\vec{S}. \quad \underset{\text{theorem}}{\overset{\text{Stokes'}}{\Longrightarrow}} \quad \nabla \times \vec{E} = -\frac{\partial}{\partial t} \vec{B}$$

tells us that a changing magnetic field generates a circulating electric field configuration:

Analogously, the **Ampere-Maxwell law** (Eq. 3.29, Eq. 3.32)

$$\oint_C \vec{B} \cdot \vec{dl} = \mu_0 \int_S \left(\vec{J} + \epsilon_0 \frac{\partial}{\partial t} \vec{E} \right) \cdot \vec{dS} \quad \xRightarrow{\text{Stokes' theorem}} \quad \nabla \times \vec{B} = \mu_0 \left(\vec{J} + \epsilon_0 \frac{\partial}{\partial t} \vec{E} \right)$$

tells us that a changing electric field generates a circulating magnetic field configuration:

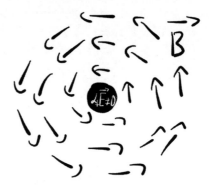

In addition, it tells us that moving electric charges (a non-zero current density $\vec{J}$) also generate a circulating magnetic field:

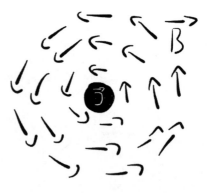

---

[73] We are oversimplifying a bit here. As we will discuss in the next chapter, only steady currents lead to a static circulating magnetic field. A single moving charged object leads to a time-dependent magnetic field configuration.

So in summary, Maxwell's equations tell us that charged objects generate specific structures in the electric and magnetic field:[73]

static charged object → static radial electric field

moving charged object → static circulating magnetic field.

Moreover, they tell us that we can also generate different structures by using the fields themselves

changing magnetic field → changing circulating electric field

changing electric field → changing circulating magnetic field.

The Lorentz force law then tells us how our charged objects react if such structures are present in our system.

In addition, we have learned that the interplay between the electric field and the magnetic field means that there can be electromagnetic waves. These are described by the wave equation for the electric field (Eq. 3.38)

$$\nabla^2 \vec{E} = \mu_0 \epsilon_0 \frac{\partial^2}{\partial t^2} \vec{E}$$

and the wave equation for the magnetic field (Eq. 3.39)

$$\nabla^2 \vec{B} = \mu_0 \epsilon_0 \frac{\partial^2}{\partial t^2} \vec{B}.$$

Now it's time to see how all this works in practice. In the following sections, we will talk about the most important electrodynamical systems and how we can describe them using the tools that we talked about in the previous two chapters.

Part II
Essential Electrodynamical Systems and Tools

"It's of no use whatsoever. This is just an experiment that proves Maestro Maxwell was right—we just have these mysterious electromagnetic waves that we cannot see with the naked eye. But they are there."

Heinrich Hertz

PS: You can discuss the content of Part II with other readers and give feedback at www.nononsensebooks.com/edyn/part2.

In the following sections, we'll discuss how we can use Maxwell's equations to describe concrete systems. An important observation is that in many systems, the electric and magnetic fields are *static*. This means that they don't change as time passes. Everything is much easier when our fields are static and therefore we will start by discussing such **electrostatic** and **magnetostatic** systems.

Afterwards, we will talk about what happens when our fields change dynamically. The most important new feature is that a changing magnetic field leads to changing electric field and vice versa. We will see that this directly implies there can be **electromagnetic waves**. An electromagnetic wave is a nontrivial structure in the electric and magnetic fields which is able to move forward as time passes.

Before we start, it is instructive to talk about one general property of solutions of Maxwell's equations and the wave equations.

An extremely important observation is that as soon as we have found the electric field configuration $\vec{E}_1$ around a specific charged object, we can calculate the total electric field configuration in more general systems in which our charged object appears by using

$$\vec{E}_{\text{sup}} = a\vec{E}_1 + b\vec{E}_2, \tag{3.41}$$

where $\vec{E}_2$ is the configuration which all other objects in the systems would generate if they were alone. This is known as a **superposition** of solutions. In other words, the total field configurations generated by various charged objects is simply the sum of the individual solutions.

Why is a superposition of solutions also a solution?

To see this, let's assume we have found two solutions $\vec{E}_1, \vec{E}_2$ of the wave equation for the electric field (Eq. 3.38).[74] Mathemati-

[74] The same reasoning is true for *any* linear equation, i.e. for Maxwell's equations and the wave equation for the magnetic field too.

cally, this means that

$$\nabla^2 \vec{E}_1 = \mu_0 \epsilon_0 \frac{\partial^2}{\partial t^2} \vec{E}_1$$

$$\nabla^2 \vec{E}_2 = \mu_0 \epsilon_0 \frac{\partial^2}{\partial t^2} \vec{E}_2. \quad (3.42)$$

Putting a superposition of these two solutions (Eq. 3.41) into the wave equation yields

$$\nabla^2 \vec{E}_{\text{sup}} = \mu_0 \epsilon_0 \frac{\partial^2}{\partial t^2} \vec{E}_{\text{sup}}$$

↝ Eq. 3.41

$$\therefore \quad \nabla^2 \left(a\vec{E}_1 + b\vec{E}_2 \right) = \mu_0 \epsilon_0 \frac{\partial^2}{\partial t^2} \left(a\vec{E}_1 + b\vec{E}_2 \right)$$

↝

$$\therefore \quad a\nabla^2 \vec{E}_1 + b\nabla^2 \vec{E}_2 = \mu_0 \epsilon_0 a \frac{\partial^2}{\partial t^2} \vec{E}_1 + \mu_0 \epsilon_0 b \frac{\partial^2}{\partial t^2} \vec{E}_2$$

↝ Eq. 3.42

$$\therefore \quad a\left(\mu_0 \epsilon_0 \frac{\partial^2}{\partial t^2} \vec{E}_1\right) + b\left(\mu_0 \epsilon_0 \frac{\partial^2}{\partial t^2} \vec{E}_2\right) = \mu_0 \epsilon_0 a \frac{\partial^2}{\partial t^2} \vec{E}_1 + \mu_0 \epsilon_0 b \frac{\partial^2}{\partial t^2} \vec{E}_2 \quad \checkmark$$

So the linear combination $\vec{E}_{\text{sup}}$ really is also a solution.[75]

[75] This only works because there is no term $\vec{E}^2$ or $\vec{E}^3$ in the wave equation. In other words, as already mentioned above, superpositions are automatically solutions only if the equation is linear. If you don't believe this, try to do the same calculation with an additional term $\vec{E}^2$ or $\vec{E}^3$ in the wave equation.

This observation is important because it allows us to investigate the implications of simple systems individually. Then we can use these results to get solutions for complicated systems by using appropriate sums of the simple solutions. In other words, we can act as if each object in the systems generates its own field and the total observable field is then simply the sum of these individual fields.

Using superpositions of solutions always works when the equation we want to solve is linear.[76] Formulated differently, whenever the equation is linear, the sum of two or more solutions yields another solution.

[76] Luckily, many important equations in physics are indeed linear. A noteworthy exception is the Einstein equation in general relativity and the Yang-Mills equations in quantum field theory.

With this in mind, we are ready to discuss electrostatics and magnetostatics.

4

Electrostatics and Magnetostatics

As in the previous sections, we will start with a birds-eye overview and only afterwards dive into the details.

In many important electromagnetic systems, the magnetic and electric field do not change over time. This is the case if the charge density and the current density are time-independent.

If the charge density is **static** ($\partial_t \rho = 0$), the resulting electric field configuration is time-independent.[1] Analogously, if our current is **steady** ($\partial_t \vec{J} = 0$), the resulting magnetic field configuration is time-independent.

In such systems we can treat the electric field and the magnetic field independently because only if the electric field is changing does it have an effect on the magnetic field and vice versa. This is what the Ampere-Maxwell law (Eq. 3.23) and Faraday's law (Eq. 3.23) tells us. Maxwell's equations then become two separate sets of equations. One describes static electric fields while the second set describes static magnetic fields.

[1] Take note that $\partial_t \rho = 0$ does not mean that our charges aren't moving. Instead, it only means that if our charges are moving, the amount that flows in is exactly equal to the amount that flows out. For example, in an ordinary wire we can also have a static charge distribution. In a (perfect) wire there is no net charge transport since all electrons move forward together such that each electron is immediately replaced by another one as soon as it moves away. Otherwise the wire would run out of charges at some point. Formulated differently, our individual charges can move around even though there is no net charge transport. However, take note that we have $\rho = 0$ in a wire since the positive charge of the protons cancels the charge of the electrons everywhere. Hence, a wire generates a nontrivial magnetic field configuration but has no influence on the electric field.

For the electric field, we have

$$\nabla \cdot \vec{E} = \frac{\rho}{\epsilon_0} \quad \text{(electric Gauss' law, Eq. 3.16)}$$

$$\nabla \times \vec{E} = 0 \quad \text{(Faraday's law, Eq. 3.27 with } \partial_t \vec{B} = 0\text{)}. \quad (4.1)$$

These are the fundamental equations of **electrostatics**. We can use these equations to calculate the electric field configuration $\vec{E}(\vec{x})$ around a given *static* charge distribution $\rho(\vec{x})$. A famous example is the electric field configuration around a charged sphere.[2]

[2] We will discuss the electric field configuration around a sphere in Section 4.1.2.

Analogously, for the magnetic field we have

$$\nabla \cdot \vec{B} = 0 \quad \text{(magnetic Gauss' law, Eq. 3.20)}$$

$$\nabla \times \vec{B} = \mu_0 \vec{J} \quad \text{(Ampere-Maxwell law, Eq. 3.32, with } \partial_t \vec{E} = 0\text{)}, \quad (4.2)$$

which are known as the fundamental equations of **magnetostatics**. We can use these equations to calculate the magnetic field configuration $\vec{B}(\vec{x})$ around *steady* electric current densities $\vec{J}(\vec{x})$. A famous example, is the magnetic field configuration around a long wire carrying a steady electric current.[3]

[3] We will discuss the magnetic field configuration around a wire in Section 4.2.1.

The general solution of the electrostatic equations (Eq. 4.1) reads

$$\boxed{\vec{E}(\vec{r}) = \frac{1}{4\pi\epsilon_0} \int \frac{\rho(\vec{r}')(\vec{r} - \vec{r}')}{|\vec{r} - \vec{r}'|^3} d^3\vec{r}',} \quad (4.3)$$

which is known as **Coulomb's law**. Here, $\vec{r}$ is a vector that points to the point we're evaluating the field at. Moreover, $\vec{r}'$ points to the location of the source of the electric field.

In words, this solution tells us that the electric field configuration generated by a general charge distribution is simply the sum over the field configurations generated by the individual charges.

We can see this by writing our charge distribution in terms of individual charges[4]

$$\rho(\vec{r}) = \sum_i q_i \delta(\vec{r} - \vec{r}_i). \quad (4.4)$$

[4] Reminder: the charge distribution for a single point charge is

$$\rho(\vec{x}) = q \delta(\vec{x} - \vec{x}'),$$

where $\vec{x}'$ is the location of the charge, q is its electric charge and $\delta(\vec{x} - \vec{x}')$ is the delta distribution. We talked about this in Section 2.2. In addition, the delta distribution is discussed in Appendix C.

Putting this into Eq. 4.3 yields

$$\vec{E}(\vec{r}) = \frac{1}{4\pi\epsilon_0} \int \frac{\rho(\vec{r}')(\vec{r}-\vec{r}')}{|\vec{r}-\vec{r}'|^3} d^3\vec{r}' \qquad \text{this is Coulomb's law, Eq. 4.3}$$

$$= \frac{1}{4\pi\epsilon_0} \int \frac{\left(\sum_i q_i \delta(\vec{r}'-\vec{r}_i)\right)(\vec{r}-\vec{r}')}{|\vec{r}-\vec{r}'|^3} d^3\vec{r}' \qquad \text{Eq. 4.4}$$

$$= \sum_i \frac{q_i}{4\pi\epsilon_0} \frac{\vec{r}-\vec{r}_i}{|\vec{r}-\vec{r}_i|^3} . \qquad (4.5) \qquad \int \delta(x-a)f(x)dx = f(a)$$

We can visualize this as follows

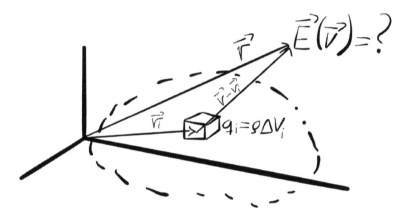

The final puzzle piece we need is that the field configuration generated by a single point charge q localized at $\vec{r}_i$ is[5]

[5] We will derive this in Section 4.1.1.

$$\vec{E}(\vec{r}) = \frac{q}{4\pi\epsilon_0} \frac{\vec{r}-\vec{r}_i}{|\vec{r}-\vec{r}_i|^3}.$$

Therefore, by looking at Eq. 4.5 again, we can conclude that Coulomb's law really tells us that the total field configuration $\vec{E}(\vec{r})$ resulting from a general charge distribution $\rho(\vec{r})$ is simply given by the sum over the field configurations of the individual charges.

Of course, this is not a very surprising result because we know already that superpositions of solutions are solutions too. Therefore, the total electric field in any given system can be understood as a sum over the field configurations generated by the individual charges.

With this general solution at hand, we can calculate the explicit form of the electric field around specific time-independent charge distributions:

system	electric field configuration
point charge q	$\vec{E} = \frac{1}{4\pi\epsilon_0} \frac{q}{r^2} \vec{e}_r$
charged sphere Q	$\vec{E} = \frac{1}{4\pi\epsilon_0} \frac{Q}{r^2} \vec{e}_r$
charged straight wire (infinitely long with charge density λ)	$\vec{E} = \frac{1}{2\pi\epsilon_0} \frac{\lambda}{r} \vec{e}_r$
charged plane (infinitely large with charge density σ)	$\vec{E} = \frac{\sigma}{2\epsilon_0} \hat{n}$
dipole (distance d between q and $-q$)	$\vec{E}(r) = \frac{q}{4\pi\epsilon_0 |\vec{r}|^2} \vec{e}_r - \frac{q}{4\pi\epsilon_0 |\vec{r}-\vec{d}|^2} \vec{e}_r$

This tells us that the electric field configurations around these charges objects look as follows:

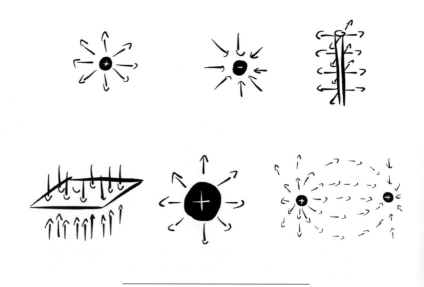

Analogously, the general solution of the magnetostatics equations (Eq. 4.2) reads

$$\vec{B}(\vec{r}) = \frac{\mu_0}{4\pi} \int \frac{\vec{J}(\vec{r}') \times (\vec{r} - \vec{r}')}{|\vec{r} - \vec{r}'|^3} d^3 r', \qquad (4.6)$$

which is known as the **Biot-Savart law**.

In words, it tells us that the magnetic field configuration generated by a general current density is simply the sum over the field configurations generated by the individual moving charges.[6] We can see this by writing our current density in terms of the movement of the individual charges[7]

$$\vec{J}(\vec{r}) = \sum_i q_i \delta(\vec{r} - \vec{r}_i) \vec{v}_i, \quad (4.7)$$

where $\vec{v}_i$ denotes the velocities of the point charges. Putting this into Eq. 4.6 yields

$$\vec{B}(\vec{r}) = \frac{\mu_0}{4\pi} \int \frac{\vec{J}(\vec{r}') \times (\vec{r} - \vec{r}')}{|\vec{r} - \vec{r}'|^3} d^3 \vec{r}' \quad \text{this is the Biot-Savart law, Eq. 4.6}$$

$$\circlearrowright \text{ Eq. 4.7}$$

$$= \frac{\mu_0}{4\pi} \int \frac{\left(\sum_i q_i \delta(\vec{r}' - \vec{r}_i) \vec{v}_i\right) \times (\vec{r} - \vec{r}')}{|\vec{r} - \vec{r}'|^3} d^3 \vec{r}'$$

$$\circlearrowright \int \delta(x-a) f(x) dx = f(a)$$

$$= \sum_i \frac{\mu_0 q_i}{4\pi} \vec{v}_i \times \frac{\vec{r} - \vec{r}_i}{|\vec{r} - \vec{r}_i|^3}. \quad (4.8)$$

We can visualize this as follows

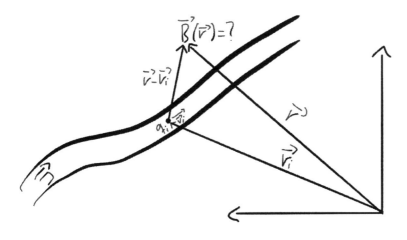

The second puzzle piece we need is that the field configuration generated by a single moving point charge q at $\vec{r}'$ moving with the velocity $\vec{v}$ is[8]

[6] This is completely analogous to what we discussed for Coulomb's law.

[7] Here, we use the definition of the charge density (Eq. 2.12):

$$\vec{J}(\vec{r}) \equiv \rho(\vec{r})\vec{v},$$

which for a single point charge q at $\vec{r}_i$ reads

$$\vec{J} \equiv q\delta(\vec{r} - \vec{r}_i)\vec{v}.$$

[8] Take note that a single moving point charge is not a magnetostatic system since it generates a time-dependent electric field configuration. However, multiple moving point charges together can yield a magnetostatic system. In other words, only lots of moving charges together can yield a steady current and therefore a magnetostatic system.

$$\vec{B}(\vec{r}) = \frac{\mu_0 q}{4\pi} \vec{v} \times \frac{\vec{r} - \vec{r}'}{|\vec{r} - \vec{r}'|^3}.$$

Therefore, by looking at Eq. 4.8 again we can see, as promised above, that the Biot-Savart law simply tells us that the total field configuration $\vec{B}(\vec{r})$ resulting from a general current density $\vec{J}(\vec{r})$ is given by the sum over the field configurations generated by the movement of the individual charges.

With this general solutions at hand, we can calculate the explicit form of the magnetic field around specific time-independent electric currents:

system	magnetic field configuration
straight wire (carrying current I)	$\vec{B} = \frac{\mu_0 I}{2\pi r} \vec{e}_\varphi$
circular loop (in xy-plane, radius R, carrying current I)	$\vec{B} = \frac{\mu_0 I R^2}{2(z^2 + R^2)^{3/2}} \vec{e}_z$
solenoid (N turns, length l, carrying current I)	$\vec{B} = \frac{\mu_0 N I}{l} \vec{e}_x$ (inside)
torus (N turns, radius R, carrying current I)	$\vec{B} = \frac{\mu_0 N I}{2\pi r} \vec{e}_\varphi$ (inside)

Therefore, the magnetic field configuration around these steady currents look as follows:

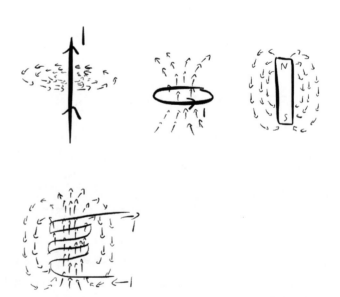

ELECTROSTATICS AND MAGNETOSTATICS

To summarize:

Electrostatics

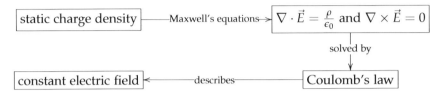

Magnetostatics

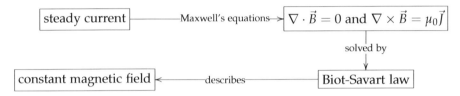

With all this in mind, we are ready to dive in and can discuss concrete electrostatic and magnetostatic systems.

4.1 Electrostatic Systems

We will start with the simplest system possible: a single point charge.

4.1.1 Electric field of a single point charge

In this example, we want to find the structure of the electric field $\vec{E}$ which is generated by a single point charge sitting at the origin of our coordinate system $\rho(\vec{r}) = q\delta(\vec{r})$.[9] Gauss's law (Eq. 3.14) tells us

[9] A point charge is an object for which the entire electric charge is localized at a single point.

$$\oint_S \vec{E} \cdot d\vec{S} = \frac{1}{\epsilon_0} \int_V \rho dV. \qquad (4.9)$$

For simplicity we choose a sphere with radius r around the point charge as our surface S:

Since we are considering a single point charge, the integral on the right-hand side simply yields q, the charge of the point charge. For Gauss's law this means:

$$\oint_S \vec{E} \cdot d\vec{S} = \frac{1}{\epsilon_0} \int_V q\delta(\vec{r}) dV$$

$$\int_S \vec{E} \cdot d\vec{S} = \frac{q}{\epsilon_0} \int_V \delta(\vec{r}) dV$$

$$\int_V \delta(\vec{r}) dV = 1$$

$$\int_S \vec{E} \cdot d\vec{S} = \frac{q}{\epsilon_0}$$

$$\int_0^{2\pi} \int_0^{\pi} \vec{E} \cdot \vec{e}_r r^2 \sin(\theta) \, d\theta \, d\phi = \frac{q}{\epsilon_0}, \tag{4.10}$$

where in the last step we switched to spherical coordinates and $\vec{e}_r$ denotes the radial unit vector which is normal to the sphere.

Our task is to solve this equation for $\vec{E}$. This is not easy and we need some clever arguments to solve this problem. The crucial idea is that since our point charge is spherically symmetric, the structure of the electric field around the point charge will be spherically symmetric too.

This is necessarily the case because our point charge does not single out any specific direction and therefore our electric field cannot point in any specific direction. In other words, since our point charge is spherically symmetric, rotating it cannot have any effect on the electric field. If the electric field around the point charge were to point in any specific direction, rotating the point charge would have a measurable effect on the electric field and this cannot be the case since the point charge is spherically symmetric.[10]

[10] Take note that the same argument applies for any spherically symmetric charge distribution, not only a point charge.

Spherical symmetry means that our electric field only depends on r, but not on θ and ϕ. Any dependence on the angles θ and ϕ would mean immediately that a rotation makes a difference and therefore our electric field wouldn't be rotationally symmetric.

So we have $\vec{E}(r,\theta,\phi) = E(r)\vec{e}_r$. In words, this means that our electric field vectors all point radially away or towards the point charge but do not, for example, circulate around the surface S.

With this in mind, we can evaluate the integral in Eq. 4.10:

$$\int_0^{2\pi} \int_0^{\pi} \vec{E} \cdot \vec{n} r^2 \sin(\theta) \, d\theta \, d\phi = \frac{q}{\epsilon_0}$$

↷ using the explicit structure of $\vec{E} = E\vec{e}_r$

$$\int_0^{2\pi} \int_0^{\pi} E \underbrace{\vec{e}_r \cdot \vec{e}_r}_{=1} r^2 \sin(\theta) \, d\theta \, d\phi = \frac{q}{\epsilon_0}$$

↷ since E does not depend on θ, ϕ

$$Er^2 \int_0^{2\pi} \int_0^{\pi} \sin(\theta) \, d\theta \, d\phi = \frac{q}{\epsilon_0}$$

↷ $\int_0^{2\pi} \int_0^{\pi} \sin(\theta) \, d\theta \, d\phi = 4\pi$

$$Er^2 4\pi = \frac{q}{\epsilon_0}$$

↷

$$E = \frac{q}{r^2 4\pi \epsilon_0} . \qquad (4.11)$$

This is the electric field strength at a distance r to the point charge q. The electric field therefore reads

$$\vec{E}(r) = \frac{q}{r^2 4\pi \epsilon_0} \vec{e}_r \qquad (4.12)$$

and looks like this (depending on the sign of the charge q)

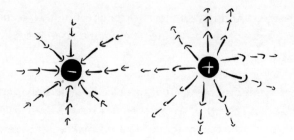

Now, we can ask: what's the force that such a single point charge exerts on another charged object?

To answer this question, we can use the Lorentz force law:

$$\vec{F} = Q\vec{E} \quad \text{Lorentz force law, Eq. 3.12 with } \vec{B} = 0$$
$$\text{Eq. 4.12}$$
$$\vec{F} = Q \frac{q}{r^2 4\pi\epsilon_0} \vec{e}_r, \tag{4.13}$$

where Q denotes the charge of the second charged object.

The final equation

$$\boxed{\vec{F}_C = \frac{Qq}{r^2 4\pi\epsilon_0} \vec{e}_r,} \tag{4.14}$$

is known as **Coulomb's law**. It describes the force which two charged objects with charges q and Q exert on each other. The force depends inversely on the distance r between the two objects. This means that the force on two charged objects close to each other is much larger than the force on two far away objects.[11]

[11] Of course, according to Newton's third law, there is an equal force on both charges. This means that in our example there is also a force $\vec{F}_C$ pointing in the opposite direction on the larger object.

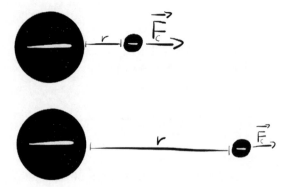

Moreover, it tells us that the force points along the axis between the two objects. The actual direction depends on the relative signs of the two charges (q and Q). If both have the *same* sign, the force is *positive*[12] which means the two objects get pushed away from each other. If the charges of the two objects have opposite signs, the force is negative and this means that the two objects attract each other.

[12] $qQ > 0$ for either $q > 0$ and $Q > 0$, i.e. if both are positively charged or $q < 0$ and $Q < 0$, i.e. if both are negatively charged.

The corresponding electric potential[13]

$$\phi = \frac{Qq}{r4\pi\epsilon_0}, \quad (4.15)$$

[13] We discussed the electric potential in Section 2.5. The potential here is the correct one which yields the force in Eq. 4.14 if we calculate the gradient $-\nabla\phi = \vec{F}_C$, where $\phi = cA_0$ and c denotes the speed of light.

is known as the **Coulomb potential**. It tells us the potential energy of an object with charge Q in the electric field of an object with charge q:

Take note that we can also calculate the correct electric field configuration by putting our charge distribution $\rho = q\delta(\vec{r})$

directly into Coulomb's general law (Eq. 4.3):

$$\vec{E}(\vec{r}) = \frac{1}{4\pi\epsilon_0} \int \frac{\rho(\vec{r}')(\vec{r}-\vec{r}')}{|\vec{r}-\vec{r}'|^3} d^3\vec{r}' \qquad \text{this is Coulomb's law, Eq. 4.3}$$

$$= \frac{1}{4\pi\epsilon_0} \int \frac{q\delta(\vec{r}')(\vec{r}-\vec{r}')}{|\vec{r}-\vec{r}'|^3} d^3\vec{r}' \qquad \curvearrowright \rho(\vec{r}') = q\delta(\vec{r}')$$

$$= \frac{1}{4\pi\epsilon_0} \frac{q\vec{r}}{|\vec{r}|^3} \qquad \curvearrowright \int dx\, \delta(x) f(x) = f(0)$$

$$= \frac{q}{4\pi\epsilon_0} \frac{|\vec{r}|\vec{e}_r}{|\vec{r}|^3} \qquad \curvearrowright \vec{r} = |\vec{r}|\vec{e}_r$$

$$= \frac{q}{4\pi\epsilon_0} \frac{\vec{e}_r}{|\vec{r}|^2}. \qquad \curvearrowright \not{\vec{r}} \qquad (4.16)$$

This is exactly the same result that we already calculated above (Eq. 4.12).

Next, we calculate the electric field configuration around a charged sphere. This is, in some sense, the logical next step after a point charge. The charge distribution is again spherically symmetric but now no longer concentrated at a single point.

4.1.2 Electric field of a sphere

We assume that we have a uniformly charged sphere and we want to calculate the electric field configuration in the region surrounding the sphere. Our charge distribution reads

$$\rho(\vec{r} - R) = \begin{cases} \rho_0, & \text{on the sphere: } |\vec{r}| = R \\ 0, & \text{otherwise}, \end{cases} \qquad (4.17)$$

where ρ_0 is a constant that denotes the charge density on the sphere.

Gauss's law (Eq. 3.14) tells us

$$\int_S \vec{E} \cdot d\vec{S} = \frac{1}{\epsilon_0} \int_V \rho dV \qquad (4.18)$$

and as our surface, we choose a sphere S around our charged sphere.

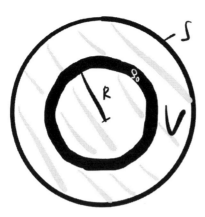

Since our sphere is uniformly charged, exactly the same arguments that we used in the previous section can be used again. Our electric field has to be spherically symmetric and this means, following exactly the same steps as in Eq. 4.11, that the result is again[14]

$$\vec{E}(\vec{r}) = \frac{q}{r^2 4\pi \epsilon_0} \vec{e}_r, \qquad (4.19)$$

where q is the total charge of the sphere.

[14] This is exactly the same result that we calculated for a single point charge in Eq. 4.12.

However, take note that we can also choose a surface S_2 which lies inside the sphere.

In this case the total charge contained within the surface is zero. Therefore Gauss's law tells us the electric field inside a charged sphere vanishes:

$$\vec{E}(\vec{r}) = 0 \qquad \text{for } |\vec{r}| < R \qquad (4.20)$$

As in the previous section, we can derive exactly the same result by using the general form of Coulomb's law (Eq. 4.3):

$$\vec{E}(\vec{r}) = \frac{1}{4\pi\epsilon_0} \int \frac{\rho(\vec{r}')(\vec{r}-\vec{r}')}{|\vec{r}-\vec{r}'|^3} d^3r' \quad \text{this is Coulombs law, Eq. 4.3}$$

$$= \frac{1}{4\pi\epsilon_0} \int \frac{\rho_0 \delta(|\vec{r}'|-R)(\vec{r}-\vec{r}')}{|\vec{r}-\vec{r}'|^3} d^3r' \quad \curvearrowright \rho(\vec{r}) \text{ in Eq. 4.17}$$

$$= \frac{\rho_0}{4\pi\epsilon_0} \int \frac{\delta(|\vec{r}'|-R)(\vec{r}-\vec{r}')}{|\vec{r}-\vec{r}'|^3} d^3r' \quad \curvearrowright$$

$$= \frac{\rho_0}{4\pi\epsilon_0} \int_0^{2\pi}\int_0^{\pi}\int_0^{\infty} \frac{\delta(|\vec{r}'|-R)(\vec{r}-\vec{r}')}{|\vec{r}-\vec{r}'|^3} r'^2 dr' \sin(\theta') d\theta' d\phi' \quad \curvearrowright$$

$$= \frac{\rho_0}{4\pi\epsilon_0} \int_0^{2\pi}\int_0^{\pi}\int_0^{\infty} \frac{\delta(|\vec{r}'|-R)(\vec{r}-\vec{r}')}{|\vec{r}-\vec{r}'|^3} r'^2 dr' \sin(\theta') d\theta' d\phi' . \quad (4.21)$$

A useful observation is that our system is spherically symmetric and hence we only need to calculate the electric field for a radial line starting at the origin. The configuration at all other points follow automatically because the electric field is spherically symmetric, i.e. we can reach all other points by rotating our solution. Therefore, for simplicity we choose $\vec{r} = (0,0,z)^T$.[15] Moreover, in general, a point in spherical coordinates is described by[16]

$$\vec{r}' = (R\sin\theta'\cos\phi', R\sin\theta'\sin\phi', R\cos\theta')^T . \quad (4.22)$$

Using this, we can calculate[17]

$$|\vec{r}-\vec{r}'| = \sqrt{(\vec{r}-\vec{r}')^2}$$

$$= \sqrt{(\vec{r})^2 + (\vec{r}')^2 - 2\vec{r}\cdot\vec{r}'}$$

$$= \sqrt{\begin{pmatrix}0\\0\\z\end{pmatrix}^2 + \begin{pmatrix}R\sin\theta'\cos\phi'\\R\sin\theta'\sin\phi'\\R\cos\theta'\end{pmatrix}^2 - 2\begin{pmatrix}0\\0\\z\end{pmatrix}\cdot\begin{pmatrix}R\sin\theta'\cos\phi'\\R\sin\theta'\sin\phi'\\R\cos\theta'\end{pmatrix}}$$

$$= \left(R^2 + z^2 - 2Rz\cos\theta'\right)^{1/2} . \quad (4.23)$$

The electric field configuration therefore reads

---

[15] Note that, as before, the superscript T denotes the transpose operation, which turns a row vector into an ordinary column vector.

[16] $\vec{r}'$ denotes the location of each charge on the sphere and therefore, we need to take all charges into account and can't restrict ourselves to a line. This is only possible for $\vec{r}$ which describes the location we want to evaluate the electric field at.

[17] We need this because this is the term which appears in the denominator of our formula.

$$\vec{E}(\vec{r}) = \frac{\rho_0}{4\pi\epsilon_0} \int_0^{2\pi} \int_0^{\pi} \int_0^R \frac{\delta(|\vec{r}'| - R)(\vec{r} - \vec{r}')}{|\vec{r} - \vec{r}'|^3} r'^2 dr' \sin(\theta') d\phi' d\theta'$$

this is Eq. 4.21

Eq. 4.23

$$= \frac{\rho_0 R^2}{4\pi\epsilon_0} \int_0^{\pi} \sin\theta' d\theta' \int_0^{2\pi} d\phi' \frac{1}{(R^2 + z^2 - 2Rz\cos\theta')^{3/2}} \begin{pmatrix} -R\sin\theta'\cos\phi' \\ -R\sin\theta'\sin\phi' \\ z - R\cos\theta' \end{pmatrix}$$

$$\int_0^{2\pi} \sin\phi \, d\phi = 0$$
$$\int_0^{2\pi} \cos\phi \, d\phi = 0$$

$$= \frac{q}{8\pi\epsilon_0} \int_0^{\pi} \sin\theta' d\theta' \frac{1}{(R^2 + z^2 - 2Rz\cos\theta')^{3/2}} \begin{pmatrix} 0 \\ 0 \\ z - R\cos\theta' \end{pmatrix}$$

$$t \equiv \cos\theta'$$
$$\Rightarrow \frac{dt}{d\theta'} = -\sin(\theta')$$

$$= \frac{q}{8\pi\epsilon_0} \int_{-1}^{1} \frac{z - Rt}{(R^2 + z^2 - 2Rzt)^{3/2}} dt \vec{e}_z$$

$$= \frac{q}{8\pi\epsilon_0} \cdot (-1) \frac{\partial}{\partial z} \int_{-1}^{1} \frac{dt}{(R^2 + z^2 - 2Rzt)^{1/2}} \vec{e}_z$$

$$= \frac{q}{8\pi\epsilon_0} \frac{\partial}{\partial z} \left[\frac{1}{Rz} \left(R^2 + z^2 - 2Rzt\right)^{1/2} \Big|_{-1}^{1} \right] \vec{e}_z$$

$$= \frac{q}{8\pi\epsilon_0} \frac{\partial}{\partial z} \left[\frac{1}{Rz} (|R - z| - |R + z|) \vec{e}_z \right]$$

$$= \frac{q}{4\pi\epsilon_0} \begin{cases} \vec{e}_z/z^2, & |z| > R \\ 0, & |z| < R \end{cases} \qquad (4.24)$$

Now, using the fact that our system is spherically symmetric, we can conclude that outside of the sphere the electric field is described by

$$\vec{E}(\vec{r}) = \frac{q}{4\pi\epsilon_0 r^2} \vec{e}_r, \qquad (4.25)$$

where $\vec{e}_r$ denotes the basis vector that points radially outward. This is exactly the same result that we derived by using Gauss's law (Eq. 4.19).

We can see that evaluating Coulomb's law can be quite complicated, even for a relatively simple system like a charged sphere.

4.1.3 Electric field of an electric dipole

An electric dipole is the logical next step after a single charged object and consists of two charges with opposite charge, i.e. q and $-q$.

For simplicity, we assume that we are dealing with two point charges and choose our coordinate system such that one charge sits at the origin. The charge density then reads

$$\rho(\vec{x}) = q\delta(\vec{x}) - q\delta(\vec{x} - \vec{d}) \tag{4.26}$$

where $\vec{d}$ is a vector pointing from the charge at the origin to the second charge.

We can now immediately write down the resulting electric field configuration because we know that we simply have to use a superposition of the individual configurations.[18]. Therefore, the total electric field configuration resulting from a dipole is the sum of the configurations generated by the two point charges independently (Eq. 4.12):

[18] We discussed this at the beginning of this part.

$$\vec{E}(r) = \vec{E}_q(r) + \vec{E}_{-q}(r) = \frac{q\vec{r}}{4\pi\epsilon_0 |\vec{r}|^3} - \frac{q(\vec{r} - \vec{d})}{4\pi\epsilon_0 |\vec{r} - \vec{d}|^3} \tag{4.27}$$

The electric field configuration therefore looks like this:

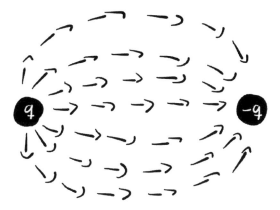

As discussed before, the mathematical reason for this result is that Maxwell's equations are *linear*.[19] To understand this point, we put our charge density into Gauss's law

[19] This means that we don't have any terms like ρ^2 or $\vec{E}^2$ etc.

$$\oint_S \vec{E} \cdot d\vec{S} = \frac{1}{\epsilon_0} \int_V \rho dV$$

$$= \frac{1}{\epsilon_0} \int_V \left(q\delta(\vec{x}) - q\delta(\vec{x} - \vec{d}) \right) dV$$

$$= \frac{1}{\epsilon_0} \int_V q\delta(\vec{x}) dV - \frac{1}{\epsilon_0} \int_V q\delta(\vec{x} - \vec{d}) dV . \quad (4.28)$$

Since there is no ρ^2 term, we simply get a sum of integrals on the right-hand side. For each of these integrals we can apply the same reasoning which we used for a single point charge in Section 4.1.1 and the result is therefore simply the sum of two point charge solutions (Eq. 4.27).

Now, after discussing systems consisting of one and two charged objects, it's time to talk about how we can make sense of systems consisting of many charged objects.

4.1.4 Electric field of more complicated charge distributions

An extremely important result is that the electric field configuration resulting from any charge distribution which is localized

[20] There is an important caveat which we will talk about in a moment.

within some region, looks exactly like the charge distribution of a single point charge if we are sufficiently far away.[20]

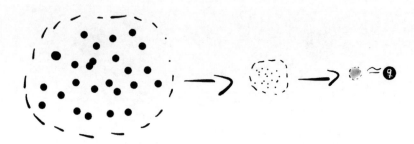

We can see this most easily by using the electric potential instead of the electric field. The general formula for the electric potential is[21]

[21] Analogous to Coulomb's law for the electric field (Eq. 4.3), this formula tells us that the total potential is the sum over the potentials generated by the individual charges. If we calculate $-\nabla \phi$ using this general form of the potential, we find exactly Coulomb's general law. We will discuss this general formula in more detail at the end of Section 4.1.6.

$$\phi(\vec{r}) = \frac{1}{4\pi\epsilon_0} \int \frac{\rho(\vec{r}')}{|\vec{r} - \vec{r}'|} d^3 r' . \qquad (4.29)$$

We assume that the charge distribution $\rho(\vec{r}')$ is localized within some region V. In mathematical terms, being far away from our charge distribution then means $|\vec{r}| \gg |\vec{r}'|$ for all $\vec{r}'$ in V.

[22] The general idea behind the Taylor expansion is discussed in Appendix B.

[23] Here $r \equiv |\vec{r}|$.

The main idea is that we can then use the Taylor expansion to simplify the general formula (Eq. 4.29).[22] In particular, we need the expansion[23]

$$\frac{1}{|\vec{r} - \vec{r}'|} = \frac{1}{r} + \vec{r}' \cdot \nabla \frac{1}{r} + \ldots$$

$$= \frac{1}{r} + \vec{r}' \cdot \frac{\vec{r}}{r^3} + \ldots \qquad (4.30)$$

Putting this expansion into our general formula (Eq. 4.29) yields

$$\phi(\vec{r}) = \frac{1}{4\pi\epsilon_0} \int \rho(\vec{r}') \frac{1}{|\vec{r}-\vec{r}'|} d^3r' \qquad \text{general formula, Eq. 4.29}$$

↷ Taylor expansion in Eq. 4.30

$$= \frac{1}{4\pi\epsilon_0} \int \rho(\vec{r}') \left(\frac{1}{r} + \vec{r}' \cdot \frac{\vec{r}}{r^3} + \ldots \right) d^3r'$$

↷ we integrate over r' not r

$$= \frac{1}{4\pi\epsilon_0 r} \int \rho(\vec{r}') d^3r'$$

↷

$$- \frac{1}{4\pi\epsilon_0 r^3} \int \rho(\vec{r}') \vec{r}' \cdot \vec{r} d^3r' + \ldots$$

The integral in the first term simply yields the total charge q contained in the volume. Therefore, if we are sufficiently far away, we can approximate our general formula as[24]

$$\phi(\vec{r}) \approx \frac{q}{4\pi\epsilon_0 r}, \qquad (4.31)$$

which is exactly the potential we found for a single point charge (Eq. 4.15). In words, this means that if we are far away, we can't distinguish between a complicated charge distribution and a simple point charge.

[24] The idea behind the Taylor expansion is that the first term in the expansion is dominant and all additional terms yield progressively small corrections. The largest correction comes from the second term etc.

However, if we look closely enough, we can observe deviations from this point charge field which are described by the higher-order terms in the Taylor expansion (Eq. 4.30). For example, if we measure our potential a bit more closely, we can describe it using the formula

$$\phi(\vec{r}) \approx \frac{q}{4\pi\epsilon_0 r} + \frac{1}{4\pi\epsilon_0 r^3} \int \rho(\vec{r}') \vec{r}' \cdot \vec{r} d^3r', \qquad (4.32)$$

It is conventional to introduce the symbol $\vec{p}$ for the quantity that characterizes this second order term:

$$\vec{p} \equiv \int \rho(\vec{r}') \vec{r}' d^3r'. \qquad (4.33)$$

Usually, $\vec{p}$ is known as the **dipole moment**. Our formula in Eq. 4.32 then reads

$$\phi(\vec{r}) \approx \frac{q}{4\pi\epsilon_0 r} + \frac{1}{4\pi\epsilon_0 r^3} \vec{p} \cdot \vec{r}. \qquad (4.34)$$

The reason for the name dipole moment is that the total charge of a dipole is zero $q - q = 0$. This means that the otherwise dominant term in Eq. 4.31 vanishes and the long distance behavior is characterized by the second term in the Taylor expansion:

$$\phi_{\text{dipole}}(\vec{r}) \approx \frac{1}{4\pi\epsilon_0 r^3} \vec{p} \cdot \vec{r}. \tag{4.35}$$

Analogously it's possible to introduce additional quantities which characterize the higher order terms in the Taylor expansion. For example, if we measure our potential extremely precisely we can use the formula

$$\phi(\vec{r}) \approx \frac{1}{4\pi\epsilon_0} \left(\frac{q}{r} + \frac{\vec{p} \cdot \vec{r}}{r^3} + \sum_{ij} \frac{Q_{ij} r_i r_j}{2r^5} \right), \tag{4.36}$$

where[25]

$$Q_{ij} \equiv \int_V \rho(\vec{r}')(3r_i' r_j' - \delta_{ij} r'^2) d^3 r' \tag{4.37}$$

is known as the **quadrupole moment**.[26]

[25] Take note that Q_{ij} has two indices and is therefore a matrix.

[26] The reason for this name is analogous to the reason for the name "dipole moment". The total charge of an electric quadrupole (four charges, $+q, +q, -q, -q$ arranged in a square) is zero and the dipole moment also vanishes. Hence the leading term in the Taylor expansion is the third one which we therefore call the quadrupole moment.

The main idea is that charge distributions which are localized within some region V, no matter how complicated, can be described using a small number of relatively simple quantities of decreasing importance:

▷ The most important feature if we look at the charge distribution from far away is the total charge q.

▷ The second-most important quantity is the dipole moment $\vec{p}$, which contains some rough information about how the charges are distributed.

▷ The third-most important quantity is the quadrupole moment Q_{ij}, which contains more detailed information about how the charges are distributed.[27]

[27] Of course, it is possible to introduce even more quantities. For example, the next term in the Taylor expansion is characterized by the so-called **octopole moment**.

This method of collecting step-by-step information about complicated charge distributions is known as **multipole expansion**.

4.1.5 Charged object in a static electric field

Now that we've learned how to derive the specific form of the electric field in various systems, it's time to talk about how electrically charged objects react to a non-zero electric field strength.

The Lorentz force law (Eq. 3.12) allows us to calculate for any given electric field configuration the resulting force $\vec{F}$ on a charged object.[28] Then, with this force at hand, we can do what we always do in classical mechanics: put it into Newton's second law $\vec{F} = m\vec{a}$. Solving this equation yields the trajectory of our charged object.

[28] Reminder:
$$\vec{F} = q(\vec{E} + \vec{v} \times \vec{B}).$$

To summarize, the steps are:

1. Determine the electric field configuration $\vec{E}(\vec{x})$ in the system.[29]

[29] We discussed in the previous sections how this works in practice.

2. Use the Lorentz force law $\vec{F} = q\vec{E}$ to determine the resulting force on a charged object.[30]

[30] This is the Lorentz force law (Eq. 3.12) for a vanishing magnetic field strength $\vec{B} = 0$.

3. Solve the differential equation given by Newton's second law $\vec{F} = m\vec{a}$ to find the trajectory of the charged object in the system.

Let's consider a simple example to see how this works.

We assume that the electric field in our system has the same strength and direction everywhere. This is, for example, the case between two charged plates:

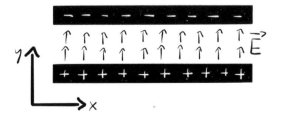

Our goal is to calculate the trajectory of a charged object with charge q which moves with some initial velocity $\vec{v}(0) = v\vec{e}_x$ perpendicular to the electric field $\vec{E} = E\vec{e}_y$. Moreover, we use a coordinate system where the charged object is exactly at the origin at $t = 0$ exactly, i.e. $\vec{x}(0) = 0$.

The Lorentz force law tells us $\vec{F} = q\vec{E} = qE\vec{e}_y$. In words, this means that there is a constant force which pushes the charge in the y-direction.

Putting this result into Newton's second law yields

$$m\vec{a} = \vec{F}$$

↷ Lorentz force law

$$m\vec{a} = qE\vec{e}_y$$

↷ $\vec{a} \equiv \frac{d^2\vec{x}}{dt^2}$

$$m\frac{d^2\vec{x}}{dt^2} = qE\vec{e}_y. \tag{4.38}$$

This is a differential equation which we need to solve using the initial condition $\vec{v}(0) = \frac{d\vec{x}}{dt}(0) = v\vec{e}_x$. The result is $\vec{x}(t)$ which describes the trajectory of the object. Since we have a constant on the right-hand side, we can integrate the equation directly

$$m\frac{d^2\vec{x}}{dt^2} = qE\vec{e}_y$$

↷ $\int dt$

$$\int dt\, m\frac{d^2\vec{x}}{dt^2} = \int dt\, qE\vec{e}_y$$

↷

$$m\frac{d\vec{x}}{dt} = qEt\vec{e}_y + \vec{c}_1$$

↷ $m\frac{d\vec{x}}{dt}(0) = mv\vec{e}_x \Rightarrow \vec{c}_1 = mv\vec{e}_x$

$$m\frac{d\vec{x}}{dt} = qEt\vec{e}_y + mv\vec{e}_x$$

↷ $\int dt$

$$\int dt\, m\frac{d\vec{x}}{dt} = \int dt\, (qEt\vec{e}_y + mv\vec{e}_x)$$

↷

$$m\vec{x} = \frac{1}{2}qEt^2\vec{e}_y + mv\vec{e}_x t + \vec{c}_2$$

↷ $\vec{x}(0) = 0 \Rightarrow \vec{c}_2 = 0$

$$m\vec{x} = \frac{1}{2}qEt^2\vec{e}_y + mv\vec{e}_x t,$$

where we determined the integration constants using the initial conditions. In words, this result tells us that our charged object will continue to move forward in the x-direction with velocity v since there is no component of the force in the x-direction. At the same time, the object gets pushed in the y-direction as a result of the electric field:

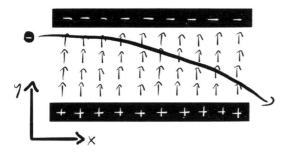

4.1.6 Further Systems

It's clear that there are an infinite number of electrostatic systems that we did not discuss here. In most of these systems nothing really new can be learned and for this reason we restricted ourselves to the fundamental systems discussed in the previous sections.[31] Nevertheless, a few comments on more advanced systems are in order.

[31] Most textbooks discuss them solely to prepare students for exams.

▷ First of all, there is a large class of electrostatic systems that we didn't discuss at all.[32] In this class of systems the positions of the charges is not fixed and known. This happens whenever there are conductors in the system where charges move around freely. Describing these kind of systems is a lot more difficult since we need the locations of the charges to calculate the resulting electric field configuration but, in turn, we need to know the field configuration in order to calculate the locations of the charges. Formulated differently, the main difficulty is that the charge density redistributes itself as a result of the electric field.

[32] If you're interested in these advanced applications, you can find excellent discussions in the textbooks recommended in Chapter 9.

The main idea is to use the so-called **uniqueness theorem**.

This theorem states that if the electric potential on the surfaces of all conductors is known, the electric potential is fixed everywhere uniquely. For this reason, finding a description of electrostatic systems where the charge distribution is not fixed and known is usually called a **boundary value problem**.

Using the uniqueness theorem it is also possible to develop the **method of image charges**, which is a clever tool to derive the electric field configuration near conductors. The trick we used here is that if we are dealing with a complicated systems, we can replace it with a simpler system with the same boundary conditions. Since the uniqueness theorem assures us that the solution of Poisson's equation is unique once the boundary conditions are fixed, we know that the simpler system and the more complicated system correspond to the same solution. The main idea is therefore to find a system with the same boundary conditions as the system we want to describe. For many systems, the same boundary conditions can be fulfilled using a few so-called image charges. Then we solve the systems consisting of our original charge and the image charges. As usual, the solution for such a system is simply the sum over the solutions of the individual charges. Since the simple system consisting of the image charges and our charge fulfills the same boundary conditions as our original system, we know immediately that the solution is also valid for our original system.

For example, let's imagine we want to calculate the electric field configuration in a system consisting of a single point charge q and an infinitely large grounded plate ($\phi = 0$ inside the plate). The presence of the point charge pushes the charge in the plate around. We don't know how exactly the resulting charge distribution on the plate looks and therefore this is quite a difficult problem. But we can solve it by using the method of image charges.

All we have to do is to find a simpler electrostatic system with the same boundary conditions, i.e. that yields $\phi = 0$ at the location of the plate. An extremely simple system which fulfills this condition consists of our original charge q and a second image charge $-q$ symmetrically behind the axis defined by the plate.[33]

[33] Take note that the plate is not present in this alternative system. As mentioned above, we replace our system consisting of (charge) + (plate) with the system consisting of (charge) + (image charge) since it is far simpler to solve.

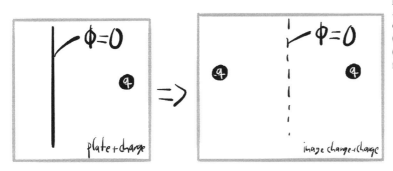

Each of the two charges in our new system generate a particular electric potential. But since we placed our image charge at a particular spot and gave it charge $-q$, the two potentials cancel exactly everywhere on the plane where the grounded plate is located in our original system.

Therefore, the two systems (charge) + (plate) and (charge) + (image charge) fulfill the same boundary conditions. The uniqueness theorem tells us that both systems have the same solution and therefore, the extremely simple solution for the system (charge) + (image charge):[34]

$$\phi(\vec{r}) = \frac{q}{4\pi\epsilon_0 |\vec{r} - d\vec{e}_x|} - \frac{q}{4\pi\epsilon_0 |\vec{r} + d\vec{e}_x|} \qquad (4.39)$$

is also a solution for the system (charge + plate). Maybe it's not obvious that our potential fulfills the boundary condition ($\phi = 0$ in the yz-plane) and we therefore check it explicitly:

$$\phi(0,y,z) = \frac{q}{4\pi\epsilon_0 |y\vec{e}_y + z\vec{e}_z - d\vec{e}_x|} - \frac{q}{4\pi\epsilon_0 |y\vec{e}_y + z\vec{e}_z + d\vec{e}_x|}$$

$$= \frac{q}{4\pi\epsilon_0 \sqrt{y^2 + z^2 + d^2}} - \frac{q}{4\pi\epsilon_0 \sqrt{y^2 + z^2 + d^2}}$$

$$= 0 \quad \checkmark$$

[34] For concreteness, we assume that the plate is in the yz-plane and our original charge is located at $\vec{r}' = (d,0,0)^T = d\vec{e}_x$. The correct location for our image charge is then $\vec{r}'' = (-d,0,0)^T$. Moreover, we use that the potential of a single point charge at $\vec{r}'$ is $\frac{q}{4\pi\epsilon_0 |\vec{r} - \vec{r}'|}$

We can then calculate the electric field configuration either by using $\vec{E} = -\nabla \phi$ or by writing down directly the superposition of the well known point charge solutions.

$$\vec{E}(\vec{r}) = \frac{q(\vec{r} - d\vec{e}_x)}{4\pi\epsilon_0 |\vec{r} - d\vec{e}_x|^3} - \frac{q(\vec{r} + d\vec{e}_x)}{4\pi\epsilon_0 |\vec{r} + d\vec{e}_x|^3} \quad (4.40)$$

▷ A common type of electrostatic problem discussed in textbooks is the electric field configuration around constant charge distributions confined to some special geometry.

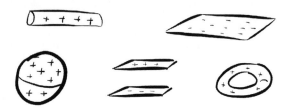

We already discussed three examples of this type of problem: a point charge, a sphere and a dipole. In general, these problems can be classified according to the dimension of the geometry which contains the non-zero charge distribution. A zero-dimensional geometry is just a point which therefore corresponds to the **point charge** example we discussed in Section 4.1.1. A one-dimensional geometry is, for example, a line and these kinds of problems therefore involve **line charges**.[35] A two-dimensional geometry is a surface and therefore contains a **surface charge** while three-dimensional charge distributions are known as **volume charges**.[36]

[35] Another popular example of this kind is a ring of charge.

[36] A popular two-dimensional example is a charged disk.

The main task in all these problems is to solve the integral in Coulomb's law (Eq. 4.3) and therefore, in some sense, this is really a mathematics problem, not a physics problem.[37]

[37] For many realistic charge distributions, Maxwell's equations can only be solved numerically anyway. Probably all examples which can be solved nicely analytically can be found in some textbook.

▷ In the context of boundary value problems it is often more convenient to use the electric potential instead of the electric field. The correct electrostatic equation for the electric potential can be derived using Maxwell's equations and the relationship between the potential ϕ and the electric field $\vec{E}$:[38]

[38] Take note that we use the conventional symbol ϕ for the electric potential and use that $\partial_t A_i = 0$ since we are dealing with a static situation.

$$\nabla \cdot \vec{E} = \frac{\rho}{\epsilon_0} \quad \text{this is the electric Gauss' law, Eq. 3.16}$$
$$\circlearrowright \; \vec{E} = -\nabla \phi \text{, c.f. Eq. 2.20 with } \partial_t A_i = 0$$
$$-\nabla^2 \phi = \frac{\rho}{\epsilon_0}. \tag{4.41}$$

This equation is known as the **Poisson equation**. If an observer is in a region external to a charge distribution where no charges are present, then $\rho = 0$ and the equation for the electric potential reads

$$-\nabla^2 \phi = 0 \quad \text{Eq. 4.41 with } \rho = 0, \tag{4.42}$$

which is known as the **Laplace equation**. Also take note that the second electrostatic equation (Eq. 4.1)

$$\nabla \times \vec{E} = 0 \quad \text{this is Eq. 4.1}$$
$$\circlearrowright \; \vec{E} = -\nabla \phi \text{, c.f. Eq. 2.20 with } \partial_t A_i = 0$$
$$-\nabla \times \nabla \phi = 0 \tag{4.43}$$

is trivially fulfilled by the potential since $\nabla \times \nabla f = 0$ ("a gradient cannot curl") holds for any scalar field f.[39]

[39] See Appendix A.16.

Take note that the Poisson equation is linear in ϕ and ρ, which means again that we can use superpositions of solutions as new solutions.

▷ There are clever methods for solving the Poisson equation. The main idea behind the most famous one is that we can calculate the solution for a general charge distribution by using the known solutions for individual charges.

[40] This is the Poisson equation (Eq. 4.41) with $\rho(\vec{r}) \sim \delta(\vec{r} - \vec{r}')$ without all constants.

Hence, the basic building block for a general solution is the solution for a single point charge. The charge distribution of a single point charge at $\vec{r}'$ is $\rho(\vec{r}) \sim \delta(\vec{r} - \vec{r}')$.

[41] In general, a Green's function characterizes the response of a system (characterized by a specific equation) to the presence of a point source.

Therefore, our main task in constructing a general solution is to solve the equation[40]

$$\nabla^2 G(\vec{r}, \vec{r}') = -\delta(\vec{r} - \vec{r}'). \tag{4.44}$$

The solution $G(\vec{r}, \vec{r}')$ is known as the **Green's function for the Laplacian operator** $\nabla^2 \equiv \Delta$.[41]

[42] We simply assume that someone somehow found this solution and we can verify that it indeed solves Eq. 4.44. The question of how to find Green's functions for a given differential operator is a difficult one and not one that we will dive into. For our modest goals it is enough to recall that we already calculated the electric field surrounding a point charge (Section 4.1.1) and can use this to derive the corresponding potential ($\vec{E} = -\nabla\phi$). The potential we find this way is exactly the Green's function we are looking for.

[43] The general idea used by the Green's function method is to find a function which yields the delta distribution if we act with the relevant differential operator (here ∇^2) on it. Then we can immediately write down the general solution like this.

The Green's function for the Laplacian operator reads[42]

$$G(\vec{r},\vec{r}') = \frac{-1}{4\pi}\frac{1}{|\vec{r}-\vec{r}'|}. \quad (4.45)$$

The main idea is that as soon as we have the solution $G(\vec{r},\vec{r}')$ for a single point charge, we can calculate directly the solution of the Poisson equation for a general charge distribution $\rho(\vec{r}')$ as follows:

$$\phi_{\text{sol}}(\vec{r}) = \frac{1}{\epsilon_0}\int \rho(\vec{r}')G(\vec{r},\vec{r}')d^3r'. \quad (4.46)$$

We can check this explicitly[43]

$$\frac{\rho(\vec{r})}{\epsilon_0} = -\nabla^2\phi_{\text{sol}}(\vec{r}) \qquad \text{this is Eq. 4.41}$$

$$\circlearrowright \text{ Eq. 4.46}$$

$$= -\nabla^2\frac{1}{\epsilon_0}\int \rho(\vec{r}')G(\vec{r},\vec{r}')d^3r'$$

$$\circlearrowright \nabla^2 \text{ acts on } \vec{r} \text{ and not on } \vec{r}'$$

$$= -\frac{1}{\epsilon_0}\int \rho(\vec{r}')\nabla^2 G(\vec{r},\vec{r}')d^3r'$$

$$\circlearrowright \text{ Eq. 4.45}$$

$$= \frac{1}{\epsilon_0}\int \rho(\vec{r}')\delta(\vec{r}-\vec{r}')d^3r'$$

$$\circlearrowright \int dx\, f(x)\delta(x-y) = f(y)$$

$$= \frac{1}{\epsilon_0}\rho(\vec{r}) \checkmark \quad (4.47)$$

Therefore, the general solution of the Poisson equation reads

$$\phi_{\text{sol}}(\vec{r}) = \frac{-1}{4\pi\epsilon_0}\int \frac{\rho(\vec{r}')}{|\vec{r}-\vec{r}'|}d^3r' \quad \text{(combining Eq. 4.45 and Eq. 4.46)}.$$

This is exactly the equation we used in Section 4.1.4 and yields Coulomb's law for the electric field (Eq. 4.3) if we use $\vec{E} = -\nabla\phi$.

Now, let's move on and talk about magnetostatic systems.

4.2 Magnetostatic Systems

As already mentioned above, a single moving point charge is not a magnetostatic system since it generates a time-dependent electric field configuration. Hence, the simplest possible magnetostatic system is instead a thin straight wire carrying a steady current.

4.2.1 Magnetic field of a wire

In this example, we want to find the structure of the magnetic field $\vec{B}$ which is generated around a wire carrying a steady current.

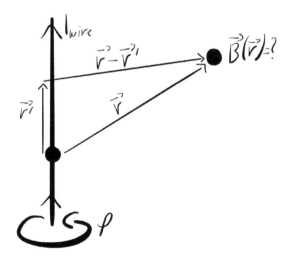

We can calculate $\vec{B}$ by using the equations of magnetostatics (Eq. 4.2). In particular, the Ampere-Maxwell law with $\partial_t E = 0$ (Eq. 3.29) tells us:[44]

$$\oint_C \vec{B} \cdot \vec{dl} = \mu_0 \int_S \vec{J} \cdot \vec{dS}. \qquad (4.48)$$

[44] Often, the Ampere-Maxwell law with $\partial_t E = 0$ is called **Ampere's law**.

To determine $\vec{B}$, our task is to get it out of the integral. The trick we can use to do this is completely analogous to what we already used in Section 4.1.1 to calculate the electric field of a single point charge. The crucial idea is that the contour C is arbitrary, as long as it encloses the wire. Hence, we choose a circle centered around the wire.

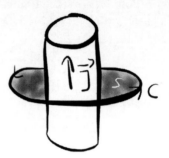

This means that our contour integral is simply an integral around a circle $\int_0^{2\pi} r\vec{e}_\varphi \, d\varphi$, where r denotes the radius of the circle and $\vec{e}_\varphi$ is a unit vector always pointing tangential when we move along the circle. Since we are dealing with a completely structureless wire, there is no reason why our magnetic field should take on different values on this circle. Moreover, from the discussion in the previous chapters, we know that our magnetic field circles around currents. Mathematically, this means that $\vec{B} = |\vec{B}|\vec{e}_\varphi$. These arguments allow us to get $|\vec{B}|$ out of the integral

$$\oint_C \vec{B} \cdot \vec{dl} = \int_0^{2\pi} |\vec{B}|\vec{e}_\varphi \cdot \vec{e}_\varphi r \, d\varphi$$

$$= |\vec{B}|r \int_0^{2\pi} d\varphi$$

$$= |\vec{B}|r2\pi. \tag{4.49}$$

On the right-hand side in Eq. 4.48, we simply get the total current through the wire I:

$$\mu_0 \int_S \vec{J} \cdot \vec{dS} = \mu_0 I_{\text{wire}}. \tag{4.50}$$

Putting these results for the left-hand and right-hand side together yields

$$\oint_C \vec{B} \cdot \vec{dl} = \mu_0 \int_S \vec{J} \cdot \vec{dS} \quad \text{this is the Ampere-Maxwell law, Eq. 4.48}$$

$$\circlearrowright \quad \text{Eq. 4.49 and Eq. 4.50}$$

$$|\vec{B}|r2\pi = \mu_0 I_{\text{wire}}$$

$$\circlearrowright$$

$$|\vec{B}| = \frac{\mu_0 I_{\text{wire}}}{r2\pi} \,. \tag{4.51}$$

The magnetic field surrounding a wire is therefore[45]

$$\vec{B}(\vec{r}) = \frac{\mu_0 q_l |\vec{v}|}{2\pi r} \vec{e}_\varphi \,. \tag{4.52}$$

[45] The wire is infinite, so the charge q on it is infinite too. Thus one really needs to work with charge per unit length which we denote by q_l.

In words, this means that the magnetic field circles around the wire. The direction in which it circles can be determined by the so-called "right hand rule". If your thumb points in the direction of the current, the remaining fingers curl in the direction of the magnetic field.

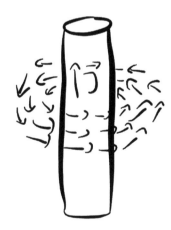

4.2.2 Charged object in a static magnetic field

Completely analogous to what we discussed in Section 4.1.5, we can calculate how a charged object reacts to the presence of a non-zero magnetic field strength. To understand how this works, we consider again a simple explicit example.

We assume that the magnetic field in our system has the same strength and direction everywhere. This is, for example, the case inside a solenoid:

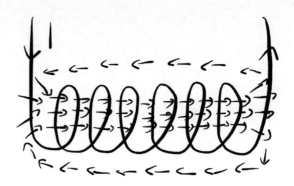

Our goal is to calculate the trajectory of a charged object with charge q which moves with some initial velocity $\vec{v}(0) = v\vec{e}_x$ perpendicular to the magnetic field $\vec{B} = B\vec{e}_z$. Moreover, for simplicity we again use a coordinate system where the charged object is at the origin at $t = 0$, i.e. $\vec{x}(0) = 0$.

The Lorentz force law tells us[46]

[46] This for a vanishing electric field strength.

$$\vec{F} = q\vec{v} \times \vec{B}$$

this is the Lorentz force law (Eq. 3.12) with $\vec{E} = \vec{0}$

$$= q \begin{pmatrix} v_x \\ v_y \\ v_z \end{pmatrix} \times \begin{pmatrix} 0 \\ 0 \\ B \end{pmatrix}$$

$$= q \begin{pmatrix} v_y B - 0 \\ 0 - v_x B \\ 0 - 0 \end{pmatrix}$$

$$= qB \begin{pmatrix} v_y \\ -v_x \\ 0 \end{pmatrix}. \qquad (4.53)$$

Putting this result into Newton's second law yields

$$m\vec{a} = \vec{F}$$

↷ $\vec{a} \equiv \frac{d\vec{v}}{dt}, \vec{F} = q\vec{v} \times \vec{B}$

$$m\frac{d\vec{v}}{dt} = q\vec{v} \times \vec{B}$$

↷ Eq. 4.53

$$m\frac{d}{dt}\begin{pmatrix} v_x \\ v_y \\ v_z \end{pmatrix} = qB\begin{pmatrix} v_y \\ -v_x \\ 0 \end{pmatrix}. \quad (4.54)$$

This differential equation is quite complicated since the velocity in the y-direction directly influences the velocity in the x-direction. However, we can disentangle it as follows[47]

[47] To unclutter the notation in the following calculations we set $m = 1$.

$$\frac{d}{dt}v_x = qBv_y \quad \text{this is the first line in Eq. 4.54}$$

↷ $\frac{d}{dt}$

$$\frac{d^2}{dt^2}v_x = qB\frac{d}{dt}v_y$$

↷ using the second line in Eq. 4.54

$$\frac{d^2}{dt^2}v_x = qB(-qBv_x)$$

↷

$$\frac{d^2}{dt^2}v_x = -q^2B^2 v_x. \quad (4.55)$$

This equation is solved by a function whose second derivative is the function itself with a minus sign. A function with exactly this property is $\sin(ct)$, where c is some appropriate constant. We can therefore conclude that our equation is solved by[48]

[48] You can also check this explicitly by putting this solution into the differential equation.

$$\vec{v}(t) = \begin{pmatrix} c_1 \cos(qBt) \\ -c_1 \sin(qBt) \\ c_2 \end{pmatrix}, \quad (4.56)$$

where c_1, c_2 are constants which we need to determine using the initial condition:

$$\vec{v}(0) = \begin{pmatrix} v \\ 0 \\ 0 \end{pmatrix}. \quad (4.57)$$

This yields

$$\vec{v}(0) = \begin{pmatrix} c_1 \cos(qB0) \\ -c_1 \sin(qB0) \\ c_2 \end{pmatrix} \stackrel{!}{=} \begin{pmatrix} v \\ 0 \\ 0 \end{pmatrix}$$

↷ $\cos(0) = 1, \sin(0) = 0$

$$\begin{pmatrix} c_1 \\ 0 \\ c_2 \end{pmatrix} \stackrel{!}{=} \begin{pmatrix} v \\ 0 \\ 0 \end{pmatrix}. \tag{4.58}$$

Our explicit solution therefore reads

$$\vec{v}(t) = \begin{pmatrix} v \cos(qBt) \\ -v \sin(qBt) \\ 0 \end{pmatrix}. \tag{4.59}$$

We can calculate the trajectory $\vec{x}(t)$ by integrating this solution

$$\vec{x}(t) = \int dt\, \vec{v}(t) = \int dt \begin{pmatrix} v \cos(qBt) \\ -v \sin(qBt) \\ 0 \end{pmatrix}$$

↷

$$= \begin{pmatrix} \frac{v}{qB} \sin(qBt) + c_3 \\ \frac{v}{qB} \cos(qBt) + c_4 \\ c_5 \end{pmatrix} \tag{4.60}$$

and using our initial condition $\vec{x}(0) = 0$:

$$\vec{x}(0) = \begin{pmatrix} \frac{v}{qB} \sin(qB0) + c_3 \\ \frac{v}{qB} \cos(qB0) + c_4 \\ c_5 \end{pmatrix} \stackrel{!}{=} \begin{pmatrix} 0 \\ 0 \\ 0 \end{pmatrix}$$

↷ $\cos(0) = 1, \sin(0) = 0$

$$\begin{pmatrix} 0 + c_3 \\ \frac{v}{qB} + c_4 \\ c_5 \end{pmatrix} \stackrel{!}{=} \begin{pmatrix} 0 \\ 0 \\ 0 \end{pmatrix}. \tag{4.61}$$

We can therefore conclude that our charged object follows the trajectory

$$\vec{x}(t) = \begin{pmatrix} \frac{v}{qB} \sin(qBt) \\ \frac{v}{qB} \cos(qBt) - \frac{v}{qB} \\ 0 \end{pmatrix}. \tag{4.62}$$

In words, this means that our charged object follows a circular path

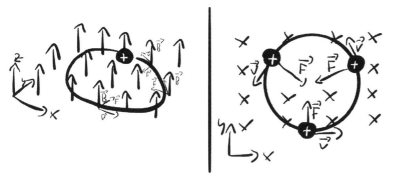

Take note that if our object's initial velocity has an additional non-zero component in the z-direction,

$$\vec{v}(0) = \begin{pmatrix} v \\ 0 \\ v_0 \end{pmatrix}, \qquad (4.63)$$

the final trajectory is a helix

$$\vec{x}(t) = \begin{pmatrix} \frac{v}{qB}\sin(qBt) \\ \frac{v}{qB}\cos(qBt) - \frac{v}{qB} \\ v_0 t \end{pmatrix}. \qquad (4.64)$$

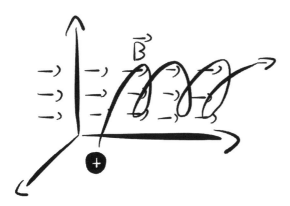

Moreover, take note that if our object initially moves *only* in the z-direction

$$\vec{v}(0) = \begin{pmatrix} 0 \\ 0 \\ v_0 \end{pmatrix}, \tag{4.65}$$

it won't be influenced by the magnetic field at all because

$$\vec{F}(0) = q\vec{v}(0) \times \vec{B} = qv_0\vec{e}_z \times B\vec{e}_z = 0 \tag{4.66}$$

since $\vec{e}_z \times \vec{e}_z = 0$, i.e. the Lorentz force is and stays zero. We can also see this because now, Eq. 4.58 reads

$$\vec{v}(0) = \begin{pmatrix} c_1 \cos(qB0) \\ -c_1 \sin(qB0) \\ c_2 \end{pmatrix} \stackrel{!}{=} \begin{pmatrix} 0 \\ 0 \\ v_0 \end{pmatrix}$$

$$\curvearrowright \cos(0) = 1, \sin(0) = 0$$

$$\begin{pmatrix} c_1 \\ 0 \\ c_2 \end{pmatrix} \stackrel{!}{=} \begin{pmatrix} 0 \\ 0 \\ v_0 \end{pmatrix}. \tag{4.67}$$

In words, this means that our object will simply continue to move in the z-direction without being influenced by the magnetic field.

4.2.3 Further Systems

There are, of course, also an infinite number of magnetostatic systems which we will not discuss here. However, a few comments are in order.

▷ Many magnetostatic systems discussed in textbooks are variations of the wire system. For example, popular examples are two parallel wires and a circular loop. In addition, it is, of course, also possible to discuss the magnetic field configuration resulting from steady currents through more complicated geometries like a solenoid, a plane or a torus.
The main task in these problems is always to solve the integral appearing in the general Biot-Savart law (Eq. 4.6).

▷ Since in magnetostatic systems we are usually dealing with thin wires, the one-dimensional analogue of the general Biot-Savart law (Eq. 4.6) is often introduced

$$\vec{B}(\vec{r}) = \frac{\mu_0}{4\pi} \int_C \frac{I\vec{dx}' \times (\vec{r} - \vec{r}')}{|\vec{r} - \vec{r}'|^3}, \qquad (4.68)$$

where I is the current and C is the path traced out by the wire. In many textbooks, this one-dimensional version is called *the* Biot-Savart law.

▷ Completely analogous to what we discussed in Section 4.1.4, it is possible to introduce **magnetic multipole moments** to characterize the long-range magnetic field configuration generated by general current densities.

▷ There is also a Poisson equation for the magnetic potential[49]

$$\nabla^2 \vec{A} = -\mu_0 \vec{J}. \qquad (4.69)$$

We have here, in fact, a Poisson equation for each component A_i. The general solution can be derived again using the Green's function method and reads[50]

$$\vec{A}_{\text{sol}}(\vec{r}) = \frac{\mu_0}{4\pi} \int \frac{\vec{J}(\vec{r}')}{|\vec{r} - \vec{r}'|} d^3 r'.$$

Then using the relationship $\vec{B} = \nabla \times \vec{A}$ between the vector potential $\vec{A}$ and the magnetic field $\vec{B}$ yields exactly the general Biot-Savart law (Eq. 4.6).

Now, let's move on and talk about electrodynamics.

[49] Take note that for the magnetic field $\nabla \cdot \vec{B} = 0$ and $\nabla \times \vec{B} = \mu_0 \vec{J}$ is the nontrivial equation. The Poisson equation for the vector potential $\vec{A}$ can then be derived by rewriting the magnetic field in terms of the potential: $\vec{B} = \nabla \times \vec{A}$ (Eq. 2.20). This yields $\mu_0 \vec{J} = \nabla \times \vec{B} = \nabla \times \nabla \times \vec{A}$. Then, we can rewrite this using the identity

$$\nabla \times \nabla \times \vec{A} = \nabla(\nabla \cdot \vec{A}) - \nabla^2 \vec{A}.$$

(See Appendix A.16.) To simplify this further, we can use the observation that potentials cannot be directly measured and only potential differences can. Hence, we can always add a constant to our potentials, and this so-called **gauge freedom** can be used to achieve $\nabla \cdot \vec{A} = 0$. If we use the gauge freedom like this, we get

$$\mu_0 \vec{J} = \nabla \times \nabla \times \vec{A}$$
$$= \nabla(\underbrace{\nabla \cdot \vec{A}}_{=0}) - \nabla^2 \vec{A}$$
$$= -\nabla^2 \vec{A},$$

which is the Poisson equation for $\vec{A}$. We will discuss gauge freedom in detail in Chapter 7.

[50] We discussed the Green's function method in Section 4.1.6.

5

Electrodynamics

In the previous chapter we focused on systems in which the electric and magnetic fields are static. Now, we are interested in what happens when the fields change dynamically. Arguably the most important new effect is the appearance of electromagnetic waves.[1]

In Section 3.7, we already used Maxwell's equations to derive the wave equations (Eq. 3.38 and Eq. 3.39)

$$\nabla^2 \vec{E} = \mu_0 \epsilon_0 \frac{\partial^2}{\partial t^2} \vec{E}$$
$$\nabla^2 \vec{B} = \mu_0 \epsilon_0 \frac{\partial^2}{\partial t^2} \vec{B}. \qquad (5.1)$$

We will now talk about explicit solutions of these equations and discuss why they describe electromagnetic waves. The main idea is that a changing magnetic field generates a changing electric field and a changing electric field generates a changing magnetic field

$$\text{changing } \vec{B} \longrightarrow \text{changing } \vec{E} \longrightarrow \text{changing } \vec{B} \longrightarrow \ldots$$

This way, nontrivial configurations in the electric and magnetic fields can travel together, for example, from the sun to the earth

[1] In fact, the role of waves in modern physics cannot be overstated. In quantum mechanics we use waves to *describe* particles. In quantum field theory, particles *are* waves. It therefore makes sense to study them in detail in a relatively simple context like electrodynamics.

[2] A nontrivial configuration is often called a perturbation of the field. The trivial or unperturbed configuration is what we call the vacuum or ground state. In physical terms, the ground state corresponds to the configuration with minimum energy, i.e. zero electric and magnetic field strengths.

although there is a vacuum in-between, i.e. no medium which could help to transmit the wave.[2]

We can understand how and why this works in practice most easily by considering an explicit solution of the wave equations.

5.1 An explicit solution of the wave equation

An explicit solution of the wave equation for the electric field $\vec{E}$ is

$$\vec{E} = A \begin{pmatrix} \cos(kz - \omega t) \\ 0 \\ 0 \end{pmatrix}, \quad (5.2)$$

where A, k and ω are constants. We can check this claim

$$\nabla^2 \vec{E} = \mu_0 \epsilon_0 \frac{\partial^2}{\partial t^2} \vec{E} \qquad \text{this is Eq. 3.38}$$

$$\nabla^2 A \begin{pmatrix} \cos(kz - \omega t) \\ 0 \\ 0 \end{pmatrix} = \mu_0 \epsilon_0 \frac{\partial^2}{\partial t^2} A \begin{pmatrix} \cos(kz - \omega t) \\ 0 \\ 0 \end{pmatrix} \qquad \text{Eq. 5.2}$$

$$-k^2 A \begin{pmatrix} \cos(kz - \omega t) \\ 0 \\ 0 \end{pmatrix} = -\mu_0 \epsilon_0 \omega^2 A \begin{pmatrix} \cos(kz - \omega t) \\ 0 \\ 0 \end{pmatrix}. \qquad \partial_z^2 \cos az = -a^2 \cos az \quad (5.3)$$

The right-hand side is equal to the left-hand side if $k^2 = \mu_0 \epsilon_0 \omega^2$. Therefore, Eq. 5.2 is a solution of the wave equation as long as this condition is fulfilled by the constants k, ω.[3]

[3] We will talk about the meaning of these constants and the condition in a moment.

A solution of this form describes a wave which travels in the z-direction with an amplitude that points in the x-direction.

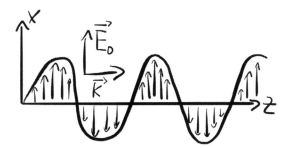

So far, we have only discussed a solution of the wave equation for the *electric* field. However, an electromagnetic wave travels through the interplay of the electric field and the magnetic field. How this comes about is the topic of the next section.

5.2 Corresponding solution of the magnetic wave equation

As already mentioned at the end of Section 3.7, the wave equation for the electric field and the wave equation for the magnetic field are only seemingly independent. Maxwell's equations must always be fulfilled. Using this fact, we can directly derive a corresponding solution of the magnetic wave equation for each solution of the electric wave equation. Hence, the two equations are not really independent, since the magnetic and electric field are still connected by Maxwell's equations.

Using our explicit solution from Section 5.1, we can understand how this works in practice. We start by recalling that Faraday's law (Eq. 3.27) describes the interplay between the electric and the magnetic field. Putting our explicit solution for the electric field $\vec{E}$ (Eq. 5.2) into Faraday's law yields

$$\nabla \times \vec{E} = -\frac{\partial}{\partial t}\vec{B} \qquad \text{this is Faraday's law, Eq. 3.27}$$

$$\downarrow \text{ Eq. 5.2}$$

$$\nabla \times A \begin{pmatrix} \cos(kz - \omega t) \\ 0 \\ 0 \end{pmatrix} = -\frac{\partial}{\partial t}\vec{B}$$

$$\downarrow \quad \nabla \times \vec{E} = \begin{pmatrix} \partial_y E_z - \partial_z E_y \\ \partial_z E_x - \partial_x E_z \\ \partial_x E_y - \partial_y E_x \end{pmatrix}$$

$$A \begin{pmatrix} 0 \\ \partial_z \cos(kz - \omega t) \\ 0 \end{pmatrix} = -\frac{\partial}{\partial t}\vec{B}$$

$$\downarrow \quad \partial_z \cos(kz) = -k\sin(kz)$$

$$A \begin{pmatrix} 0 \\ -k\sin(kz - \omega t) \\ 0 \end{pmatrix} = -\frac{\partial}{\partial t}\vec{B} .$$

(5.4)

We can conclude that the corresponding magnetic field configuration $\vec{B}$ has no component in the x-direction and no z-component: $B_x = 0, B_z = 0$. Moreover, we can calculate the y

component directly since Eq. 5.4 tells us

$$\frac{\partial}{\partial t} B_y = Ak\sin(kz - \omega t). \quad (5.5)$$

This equation is solved by

$$B_y = A\frac{k}{\omega}\cos(kz - \omega t), \quad (5.6)$$

because

$$\frac{\partial}{\partial t} A\frac{k}{\omega}\cos(kz - \omega t) = Ak\sin(kz - \omega t). \quad (5.7)$$

Therefore, if the electric field is in the configuration described by Eq. 5.2, we automatically get the magnetic field configuration:[4]

$$\vec{B} = A\frac{k}{\omega}\begin{pmatrix} 0 \\ \cos(kz - \omega t) \\ 0 \end{pmatrix}. \quad (5.8)$$

This solution describes a wave which travels in the z-direction with its amplitude pointing in the y-direction.

[4] Take note that the solution we found this way has exactly the same structure as our solution for $\vec{E}$ and is also a solution of the wave equation for the magnetic field (Eq. 3.39). We can see this immediately by putting it into the wave equation for $\vec{B}$ since the wave equation for $\vec{B}$ is completely analogous to the wave equation for $\vec{E}$. Therefore, a solution of the $\vec{E}$ wave equation automatically yields a solution for the $\vec{B}$ via Maxwell's equations.

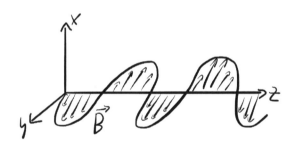

Therefore, we can conclude that an $\vec{E}$ wave which travels in the z-direction and is polarized in the x-direction (Eq. 5.2), is *automatically* accompanied by a $\vec{B}$ wave which also travels in the z-direction but which is polarized in the y-direction (Eq. 5.8):

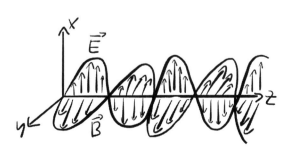

In general, the electric field and the magnetic field oscillate in phase but in perpendicular directions.

Next, after this explicit example, let's talk about solutions of the wave equations in more general terms.

5.3 General solutions of the wave equations

We will restrict our discussion to solutions of the wave equation for the electric field (Eq. 3.38) since all statements for solutions of the wave equation for the magnetic field (Eq. 3.39) follow automatically since the two equations are equivalent.

The most basic solutions of the electric wave equations look like this[5]

$$\vec{E} = \vec{E}_0 \cos(\vec{k} \cdot \vec{r} \pm \omega t + \delta), \qquad (5.9)$$

where $\vec{k}$ is a vector that determines in which direction the wave is traveling and δ encodes possible phase shifts which can be important if we are dealing with superpositions of waves.

Moreover, $\vec{E}_0$ is a vector whose magnitude $|\vec{E}_0|$ describes the amplitude of the wave and its direction determines the direction in which the electric field oscillates. The choice of the sign $\pm$ in the cosine function determines whether the wave travels in the positive direction or negative direction on the axis defined by $\vec{k}$, e.g. on the x-axis to the right or to the left. The constant ω describes the frequency at which the wave oscillates and $\vec{k}$ is directly related to the wavelength.[6]

[5] Completely analogous to what we did in the previous sections, you can check that this is indeed a solution by putting it into the wave equation (Eq. 3.38). Moreover, take note that $\delta = \pi/2$ means that our cosine function becomes a sine function. In other words, there is nothing special about our choice of the cosine function and we could equally use $\sin(\vec{k} \cdot \vec{r} \pm \omega t + \delta)$.

[6] Below, we will talk about these interpretations in more detail.

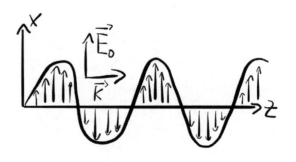

The solution discussed in the previous sections therefore corresponds to the special choice

$$\vec{E}_0 = A\vec{e}_x, \quad \vec{k} = k\vec{e}_z, \quad \delta = 0, \quad \pm \to -, \quad (5.10)$$

since

$$\vec{E} = \vec{E}_0 \cos(\vec{k} \cdot \vec{r} \pm \omega t + \delta) \qquad \text{this is Eq. 5.9}$$

$$= A\vec{e}_x \cos(k\vec{e}_z \cdot \vec{r} - \omega t) \qquad \text{Eq. 5.10)}$$

$$= A \begin{pmatrix} 1 \\ 0 \\ 0 \end{pmatrix} \cos\left(\begin{pmatrix} 0 \\ 0 \\ k \end{pmatrix} \cdot \begin{pmatrix} x \\ y \\ z \end{pmatrix} - \omega t \right)$$

$$= A \begin{pmatrix} \cos(kz - \omega t) \\ 0 \\ 0 \end{pmatrix}. \qquad (5.11)$$

This is exactly our explicit solution in Eq. 5.2.

Solutions of the form in Eq. 5.9 are known as **plane wave solutions** and are the basic building blocks for all possible solutions of the wave equations. In physical terms these solutions describe *monochromatic* electromagnetic waves.[7]

The wave equations are, like Maxwell's equations, linear in $\vec{E}$ and $\vec{B}$. Therefore, we can use superpositions of known solutions as additional solutions.[8]

This means that we can use sums of the basic plane wave solutions to create more complicated solutions. A simple example is

[7] For example, for $\omega = 60 \times 10^{12} \frac{\text{rad}}{\text{s}}$ we are dealing with what we usually call red light. ω is the angular frequency which is directly related to the more familiar ordinary frequency f: $\omega = 2\pi f$. However, take note that pure plane waves do not exist in nature. A plane wave spreads out over all space with equal amplitude. This means that an infinite amount of energy is required to create such a plane wave and is therefore impossible. Realistic solutions are wave packets, which are localized within some finite region and can be created with a finite amount of energy. "Monochromatic" light in experiments is never completely monochromatic but still contains some range of wavelengths. We will talk about wave packets below.

[8] As mentioned before, this principle of superposition is true for all linear equations.

a **standing wave**

$$\vec{E} = \vec{E}_0\Big(\cos(\vec{k}\cdot\vec{r} - \omega t) + \cos(\vec{k}\cdot\vec{r} + \omega t)\Big). \quad (5.12)$$

A solution of this type describes a superposition of a wave with a second wave with equal amplitude which travels in the *opposite* direction. The resulting total wave is not traveling in any direction but still oscillating.

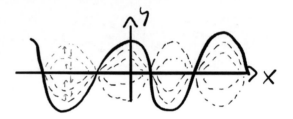

To see this mathematically, we use a trigonometric identity

$$\vec{E} = \vec{E}_0\Big(\cos(\vec{k}\cdot\vec{r} - \omega t) + \cos(\vec{k}\cdot\vec{r} + \omega t)\Big)$$

this is Eq. 5.12
$\cos(a+b) + \cos(a-b) = 2\cos(a)\cos(b)$

$$= 2\vec{E}_0\Big(\cos(\vec{k}\cdot\vec{r})\cos(\omega t)\Big). \quad (5.13)$$

The crucial point is that the spatial dependence $\vec{k}\cdot\vec{r}$ of the wave is completely separated from the time dependence ωt. This means that the spatial shape of the wave is fixed and does not change as time moves on. However, since we are still multiplying this fixed form of the wave by $\cos(\omega t)$, the *amplitude* of the wave at each point still varies (but not the wave form).

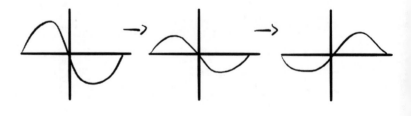

In general we can, of course, also construct arbitrarily complicated linear combinations of plane waves. A famous example is sunlight which consists of all kinds of different monochromatic light. We can see this by using a prism. Waves of different wavelength get diffracted differently and as a result the different monochromatic waves are spatially separated by the prism.

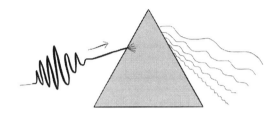

In particular, the most general solution of the wave equation is a linear combination of all possible plane waves[9]

$$\vec{E}(\vec{r},t) = \sum_{k_x=-\infty}^{\infty} \sum_{k_y=-\infty}^{\infty} \sum_{k_z=-\infty}^{\infty} \vec{E}_{\vec{k}} \cos\left(\vec{k}\cdot\vec{r} - \omega t\right). \quad (5.14)$$

[9] We use the notation $\vec{k} = (k_x, k_y, k_z)^T$. Moreover, take note that solutions with a plus in the cosine function are also included since $\vec{k}$ also takes on negative values and we can use $\cos(-x) = \cos(x)$.

Each possible wave appears in this general solution with a particular amplitude $\vec{E}_{\vec{k}}$. In general, there is no reason why our waves should only appear with a discrete set of wavelengths and therefore, we have to replace the sum with an integral[10]

$$\vec{E}(\vec{r},t) = \int_{-\infty}^{\infty} \frac{d^3k}{(2\pi)^3} \vec{E}(\vec{k}) \cos\left(\vec{k}\cdot\vec{r} - \omega t\right). \quad (5.15)$$

[10] Now that our variables k_x, k_y, k_z take on continuous values, our discrete set of amplitudes $\vec{E}_{\vec{k}}$ becomes a function $\vec{E}(\vec{k})$. However, the meaning is still the same: each possible plane wave appears in this general solution with a particular amplitude. Moreover, $(2\pi)^3$ is a conventional normalization factor.

By using such a linear combination of plane wave solutions it's possible to construct any waveform you can imagine. An extremely important example are **wave packets**.

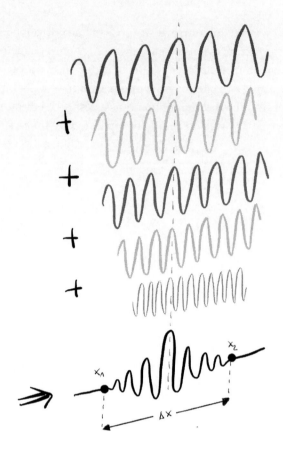

While a plane wave spreads out all over space with equal amplitude, a wave packet is localized within some finite region and can be created using a finite amount of energy.

Now, as promised above, we will discuss the properties of electromagnetic waves in a bit more detail.

5.4 Basic properties of electromagnetic waves

First of all, let's talk about the various quantities $(\vec{k}, \omega, \pm, \vec{E}_0, \delta)$ which appear in solutions of the wave equation (Eq. 5.9) in a bit more detail.[11]

[11] Reminder: Eq. 5.9 reads
$$\vec{E} = \vec{E}_0 \cos(\vec{k} \cdot \vec{r} \pm \omega t + \delta),$$

▷ The argument $\varphi \equiv (\vec{k} \cdot \vec{r} \pm \omega t)$ of our periodic function $\cos \varphi$ is called the **phase** of the wave. A phase of zero $\varphi = 0$ means we are at the top of our waveform since $\cos(0) = 1$. If the phase is $\varphi = \pi$, we are at the bottom since $\cos(\pi) = -1$.

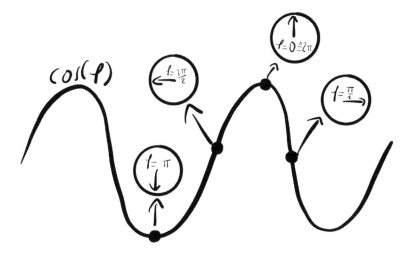

One full wave cycle starts at $\varphi = 0$ and goes on until $\varphi = 2\pi$. A periodic wave repeats itself after 2π since $\cos(\varphi + 2\pi) = \cos(\varphi)$.[12]

[12] Take note that it is conventional in theoretical physics to measure angles in multiplies of π, e.g. $180° \leftrightarrow \pi$ and for a full circle $360° \leftrightarrow 2\pi$.

▷ The vector $\vec{k}$ is usually called the **wave vector**. The direction of $\vec{k}$ tells us in which direction the wave is traveling. For example, we have seen above (Eq. 5.11) that for $\vec{k} = (0, 0, k)^T = k\vec{e}_z$, we are getting a wave which travels in the z-direction.

The length of the wave vector $|\vec{k}|$ describes how many oscillations there are *per meter*. To understand this, imagine that we could stop the time, i.e. keep t fixed and then move through space. As we move along the axis defined by $\vec{k}$ we count how many full wave shapes we encounter per meter. This number is the wave number. One full oscillation is over as soon as the phase of the wave has increased by 2π. Hence, we can say a bit more precisely that $|\vec{k}|$ measures how many 2π cycles there are per meter.[13] For this reason, $|\vec{k}|$ is known as spatial angular frequency or **wave number**.[14]

For example, if we move 2 meters and observe that the phase changes by 20π, we know that the wave number is 10π radians per meter.[15]

The wave number is directly related to the **wavelength** λ:

$$\lambda = \frac{2\pi}{|\vec{k}|}. \qquad (5.16)$$

The wavelength is defined as the spatial distance that we need to move until the phase of the wave has changed by 2π:[16]

[13] Formulated differently, $|\vec{k}|$ tells us how much the phase changes as we move one meter along the wave at one fixed point in time.

[14] This is in contrast to the *temporal* angular frequency ω, which tells us how many oscillations there are per second.

[15] $20\pi / 2\text{ m} = 10\pi /\text{ m}$

[16] Formulated differently, the wavelength is the distance between adjacent identical parts of the wave.

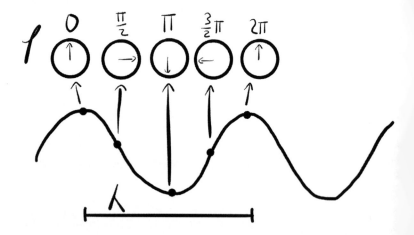

▷ The constant ω is known as temporal angular frequency or simply **angular frequency**. The angular frequency describes how many oscillations there are *per second*. To understand it, imagine that we are at one fixed point in space and time moves on. We now observe how the wave moves up and down at this one particular point. We count how often it undergoes a full oscillation, i.e. from maximum to maximum.

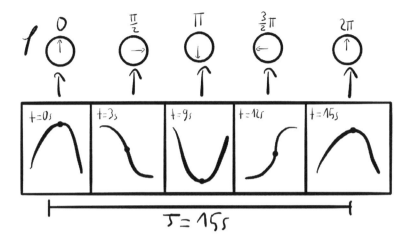

The result is the angular frequency of the wave. Formulated differently, ω tells us how much the phase changes during one second. For example, if we observe the wave for two seconds and the phase changes by 20π, i.e. we observe 10 full wave cycles, we know the angular frequency is 10π radians per second.

The angular frequency is directly related to the **period** τ and ordinary **frequency** f of the wave

$$\omega = \frac{2\pi}{\tau} = 2\pi f. \qquad (5.17)$$

The period τ is the time the wave needs for one full oscillation.

▷ The direction in which the **amplitude vector** $\vec{E}_0$ points, specifies the geometrical orientation of the oscillation.[17] Usually, the direction of $\vec{E}_0$ is called the **polarization** of the wave.

[17] Take note that this is a different direction than the direction $\vec{k}$ in which the wave travels.

The length of the amplitude vector $|\vec{E}_0|$ encodes the **peak magnitude of the oscillation**:

An extremely important observation is that the amplitude vector cannot point in any arbitrary direction for electromagnetic waves. We can see this since Gauss's law for the electric field for a system with a vanishing charge density reads

$$\nabla \cdot \vec{E} = 0.$$

This means, for example, that our solution discussed in Section 5.1 which travels in the z-direction $\vec{E} = \vec{E}(z,t)$ cannot have an oscillating amplitude in the z-direction since $\frac{\partial}{\partial z} E_z = 0$. Formulated differently, a wave which travels in the z-direction depends only on z. However, $\frac{\partial}{\partial z} E_z = 0$ then implies that $E_z = 0$. Analogously, a wave which travels in the x direction $\vec{E} = \vec{E}(x,t)$ cannot have an amplitude in the x-direction and a wave which travels in the y direction $\vec{E} = \vec{E}(y,t)$ cannot have an amplitude in the y-direction.

This means that electromagnetic waves cannot be **polarized** in the direction they are traveling. In other words, electromagnetic waves are never longitudinally polarized (at least in a vacuum). One possibility to remember this is to use the experimental result that the speed of light is an absolute speed limit in nature.[18] Nothing can move faster. Since an electromagnetic wave already travels at the speed of light, no component can travel in the direction the wave is traveling, since otherwise this part of the wave would travel with a velocity higher than the speed of light.

[18] This experimental fact is the basis for special relativity. We will talk about this in Chapter 6.

By using superpositions of plane waves, it is possible to construct waves with more complicated polarizations. For example, a linear combination of the form

$$\vec{E} = A \left(\cos(kz - \omega t)\vec{e}_x + \sin(kz - \omega t)\vec{e}_y \right), \quad (5.18)$$

describes a **circularly polarized wave**. Our amplitude vector $\vec{E}_0$ circles in the xy-plane while the plane travels in the z-direction.[19]

[19] Reminder: a circular motion in physics is described by

$$\vec{r} = \begin{pmatrix} \cos(\omega t) \\ \sin(\omega t) \end{pmatrix}.$$

In contrast, a linearly polarized wave looks like this

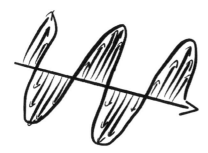

▷ The sign between the two terms in the cosine function determines whether our wave moves up or down on the axis defined by $\vec{k}$. For example, a solution of the form $\vec{E} = \vec{E}(x - ct)$ describes a wave which moves to the right on our x-axis, while a solution of the form $\vec{E} = \vec{E}(x + ct)$ describes a wave which moves to the left. This interpretation comes about since if we focus on a fixed point in our wave shape $\vec{E}(x - ct)$ and t increases, x also has to increase in order to keep $\vec{E}(x - ct)$ at the same value. In other words, this means that if we focus on a specific point in our wave shape, at a later point in time (a larger t), we will find it at a larger x.

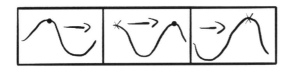

Analogously, a solution of the form $E = E(z - ct)$ describes a wave that moves up on our z-axis. Hence the solution, we discussed in Section 5.1 describes a solution which moves in the positive direction on the z-axis.

▷ The **absolute phase** δ encodes the phase of the wave at $\vec{r} = 0$ and $t = 0$.

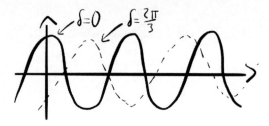

This quantity isn't measurable since it depends on how we choose our coordinate system. However, it is still important if we consider superposition of waves. If we add two waves their relative absolute phase crucially determines if the amplitude of the resulting wave is larger

in which case we speak of **constructive interference** or smaller

in which case we speak of **destructive interference**.

Now that we have a rudimentary understanding of the basic quantities associated with electromagnetic waves, we can talk about the most important of the more advanced properties.

5.5 Advanced properties of electromagnetic waves

▷ An extremely important question we haven't answered so far is: How fast are electromagnetic waves traveling?
Let's focus on a specific point on our wave. In particular, to determine the velocity of the wave, we follow the movement of a point.

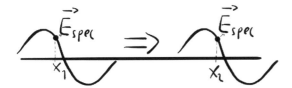

To simplify the discussion, we restrict ourselves to a wave that moves in one-dimension. With this in mind, we can calculate the velocity by using our basic solution (Eq. 5.9)

$$\vec{E} = \vec{A} \cos \left(kx - \omega t \right) = \vec{E}_0 \cos \left(k(x - \frac{\omega}{k} t) \right). \tag{5.19}$$

We assume that the specific point in our waveform $\vec{E}_{\text{spec}}$ we are interested in is at $t = t_1$ at $x = x_1$:

$$\vec{E}_{\text{spec}} = \vec{A} \cos \left(k(x_1 - \frac{\omega}{k} t_1) \right). \tag{5.20}$$

At a later point in time $t = t_2$, we will find our specific point $\vec{E}_{\text{spec}}$ at some new location $x = x_2$:

$$\vec{E}_{\text{spec}} = \vec{A} \cos \left(k(x_2 - \frac{\omega}{k} t_2) \right). \tag{5.21}$$

This means that in the interval $\Delta t = t_2 - t_1$ our specific point has traveled the distance $\Delta x = x_2 - x_1$. Therefore, our point travels with velocity

$$v = \frac{\Delta x}{\Delta t} = \frac{x_2 - x_1}{t_2 - t_1}. \tag{5.22}$$

Since we are considering the same specific point $\vec{E}_{\text{spec}}$ in Eq. 5.20 and Eq. 5.21, we can conclude

$$x_2 - \frac{\omega}{k} t_2 = x_1 - \frac{\omega}{k} t_1$$

$$\frac{\omega}{k} = \frac{x_2 - x_1}{t_2 - t_1}. \quad (5.23)$$

By comparing Eq. 5.22 with Eq. 5.23, we can conclude

$$v = \frac{\omega}{k}. \quad (5.24)$$

In words, the velocity of each point in our wave form is given by the ratio of the angular frequency ω and the wave number k.

We can also understand this from a different perspective. A velocity has units meter per second. The only combination of our basic wave quantities discussed in the previous section with units meter per second is[20]

$$v = \frac{\lambda}{\tau} \quad (5.25)$$

since the wavelength λ is measured in meters and the period in seconds. In words, this equation tells us that a wave travels one wavelength λ per period τ. We can rewrite the velocity of the wave v in terms of the angular frequency ω and wave number $k \equiv |\vec{k}|$ as follows

$$v = \frac{\lambda}{\tau} = \frac{2\pi/k}{2\pi/\omega} = \frac{\omega}{k}. \quad (5.26)$$

This is exactly the equation, we already derived above (Eq. 5.24).

▷ The formula $v = \omega/k$ is correct for waves in general. However, we need to remember that we are dealing with electromagnetic waves which are described by the wave equation (Eq. 3.38). Putting the general form of the solution (Eq. 5.9) into the wave equation yields[21]

$$k^2 = \mu_0 \epsilon_0 \omega^2. \quad (5.27)$$

This equation is known as the **dispersion relation** and tells us how k and ω have to be related if our general solution (Eq. 5.9) describes electromagnetic waves.

---

[20] To be a bit more precise: the velocity we talk about here is the **phase velocity**. This name is used to make clear that it's also possible to associate a different kind of velocity, called **group velocity**, to wave packets. The group velocity is the speed at which the envelope moves forward, while the phase velocity is the speed of the individual plane waves inside the packet. The group velocity and phase velocity are not always the same.

[21] We saw this in Section 5.1. Formulated differently, our general solution only fulfills the wave equation if this condition is satisfied.

An extremely important observation is that we can rewrite Eq. 5.27 as follows

$$k^2 = \mu_0 \epsilon_0 \omega^2 \qquad \text{Eq. 5.27}$$

$$\downarrow$$

$$\frac{k^2}{\omega^2} = \mu_0 \epsilon_0$$

$$\downarrow$$

$$\frac{\omega}{k} = \sqrt{\frac{1}{\mu_0 \epsilon_0}}. \qquad (5.28)$$

This is interesting because the ratio $\frac{\omega}{k}$ is precisely the velocity of our wave (Eq. 5.26). Therefore, the velocity of electromagnetic waves is not some arbitrary number but actually a fixed number which can be determined through the constants μ_0 and ϵ_0!

Putting in the experimental values for μ_0 and ϵ_0 yields

$$v = \sqrt{\frac{1}{\mu_0 \epsilon_0}}$$

$$\downarrow$$

$$= \sqrt{\frac{1}{\left(4\pi \times 10^{-7} \text{ m kg}/\text{C}^2\right)\left(8.854 \times 10^{-12} \text{ C}^2 \text{ s}^2/\text{kg m}^3\right)}}$$

$$\downarrow$$

$$= 2.9979 \times 10^8 \text{ m}/\text{s}. \qquad (5.29)$$

This is the velocity at which electromagnetic waves propagate. Usually, we call this number the **speed of light** and denote it by

$$c \equiv 2.9979 \times 10^8 \text{ m}/\text{s}. \qquad (5.30)$$

▷ An extremely important property of electromagnetic waves is that they can transport energy. Only thanks to this fact is life on earth possible since most of the energy we use is or was transported from the sun to earth via electromagnetic waves. To understand how this works, we first need to calculate the energy stored in a particular electric and magnetic field configuration.[22] Let's imagine that we have a specific charge distribution and current density at time t. After some small

[22] Be warned that the following derivation is somewhat indirect. Nevertheless, the final result is quite intuitive and worth the hassle.

[23] We can either imagine that all charges move with the same velocity or that $\vec{v}$ describes the average velocity.

[24] Reminder: work is defined as force times the path during which the force was applied. Moreover, take note that the magnetic force is always perpendicular to the velocity and therefore, no work is done by the magnetic field.

period dt, the charges have moved around a bit $d\vec{l} = \vec{v}dt$, where $\vec{v}$ describes the velocity of the charges.[23] Using the Lorentz force law, we can then calculate that the work done by the electromagnetic field on a single charge q is [24]

$$\vec{F} \cdot d\vec{l} = q(\vec{E} + \vec{v} \times \vec{B}) \cdot \vec{v}dt = q\vec{E} \cdot \vec{v}dt. \quad (5.31)$$

To calculate the work done on all charges, we use $q = \rho dV$ and $\vec{J} = \rho\vec{v}$. The rate at which work is done on all charges is therefore

$$\frac{dW}{dt} = \int_V (\vec{E} \cdot \vec{J}) dV. \quad (5.32)$$

The lesson to take away is that we can interpret $\vec{E} \cdot \vec{J}$ as the work done per unit time, per unit volume. In other words, $\vec{E} \cdot \vec{J}$ is the **power** per unit volume.

Next, we are interested in the energy stored in electromagnetic waves which can be present even though there are no charges and no current density $\vec{J}$ in the system. To derive a formula for such a situation, we use Maxwell's equations and the product rule for the divergence operator:

$$\begin{aligned}
\vec{E} \cdot \vec{J} &= \frac{1}{\mu_0} \vec{E} \cdot (\vec{\nabla} \times \vec{B}) - \epsilon_0 \vec{E} \cdot \frac{\partial \vec{E}}{\partial t} & \text{this is the Ampere-Maxwell law, Eq. 3.32} \\
&= \frac{1}{\mu_0} \left(\vec{B} \cdot (\vec{\nabla} \times \vec{E}) - \vec{\nabla} \cdot (\vec{E} \times \vec{B}) \right) - \epsilon_0 \vec{E} \cdot \frac{\partial \vec{E}}{\partial t} & \vec{\nabla} \cdot (\vec{E} \times \vec{B}) = \vec{B} \cdot (\vec{\nabla} \times \vec{E}) - \vec{E} \cdot (\vec{\nabla} \times \vec{B}) \\
&= \frac{1}{\mu_0} \vec{B} \cdot \left(-\frac{\partial}{\partial t} \vec{B} \right) - \frac{1}{\mu_0} \vec{\nabla} \cdot (\vec{E} \times \vec{B}) - \epsilon_0 \vec{E} \cdot \frac{\partial \vec{E}}{\partial t} & \text{Faraday's law, Eq. 3.27, } \vec{\nabla} \times \vec{E} = -\partial \vec{B}/\partial t \\
&= -\frac{1}{2\mu_0} \frac{\partial}{\partial t} \vec{B}^2 - \frac{1}{\mu_0} \vec{\nabla} \cdot (\vec{E} \times \vec{B}) - \epsilon_0 \frac{1}{2} \frac{\partial}{\partial t} \vec{E}^2. & \frac{\partial}{\partial t} \vec{B}^2 = \left(\frac{\partial}{\partial t} \vec{B} \right) \cdot \vec{B} + \vec{B} \cdot \left(\frac{\partial}{\partial t} \vec{B} \right) = 2\vec{B} \cdot \frac{\partial}{\partial t} \vec{B}
\end{aligned}$$

(5.?)

Putting this into Eq. 5.32 yields

$$\frac{dW}{dt} = \int_V (\vec{E} \cdot \vec{J}) dV \qquad \text{this is Eq. 5.32}$$

$$= \int_V \left(-\frac{1}{2\mu_0} \frac{\partial}{\partial t} \vec{B}^2 - \frac{1}{\mu_0} \nabla \cdot (\vec{E} \times \vec{B}) - \epsilon_0 \frac{1}{2} \frac{\partial}{\partial t} \vec{E}^2 \right) dV \qquad \text{Eq. 5.33}$$

↷ reorganizing the terms

$$= -\frac{1}{2} \frac{\partial}{\partial t} \int_V \left(\frac{1}{\mu_0} \vec{B}^2 + \epsilon_0 \vec{E}^2 \right) dV - \int_V \frac{1}{\mu_0} \nabla \cdot (\vec{E} \times \vec{B}) dV$$

↷ Gauss's theorem, $\int_V \nabla \cdot \vec{A} = \oint_S \vec{A}$

$$= -\frac{1}{2} \frac{\partial}{\partial t} \int_V \left(\frac{1}{\mu_0} \vec{B}^2 + \epsilon_0 \vec{E}^2 \right) dV - \oint_S \frac{1}{\mu_0} (\vec{E} \times \vec{B}) \cdot d\vec{a}$$

↷ definitions

$$\equiv -\frac{dU_{em}}{dt} - \oint_S \vec{S} \cdot d\vec{a}. \qquad (5.34)$$

This equation is known as the **Poynting theorem**.

$$\vec{S} \equiv \frac{1}{\mu_0} (\vec{E} \times \vec{B}) \qquad (5.35)$$

is usually called the **Poynting vector** and describes the energy flux per unit time through a given surface, i.e. the energy flux. Moreover, U_{em} is the total energy stored in the electromagnetic field configuration.

In words, Eq. 5.34 therefore tells us that the work done on charges by the Lorentz force is equal to the decrease in energy stored in the field $\frac{dU_{em}}{dt}$ minus the energy which has flowed through the surface S. In other words, if the energy stored in the electromagnetic field gets smaller, it must have either been used to move charges around ($\frac{dW}{dt}$) or has flown out of the volume we are considering ($\oint_S \vec{S} \cdot d\vec{a}$). Therefore, Eq. 5.34 is quite similar to the continuity equation (Eq. 3.8). However, instead of the conservation of electric charge, Eq. 5.34 describes the conservation of energy.

We can make the analogy even more concrete by deriving the differential form of Eq. 5.34. To do this, we introduce the mechanical energy density

$$\frac{dW}{dt} \equiv \frac{d}{dt} \int_V u_{mech} dV \qquad (5.36)$$

and the electromagnetic energy density

$$u_{\text{em}} \equiv \frac{1}{2}\left(\epsilon_0 \vec{E}^2 + \frac{1}{\mu_0}\vec{B}^2\right). \quad (5.37)$$

Using these definitions, we can rewrite Eq. 5.34 as follows

$$\frac{dW}{dt} = -\frac{dU_{\text{em}}}{dt} - \oint_S \vec{S}\cdot d\vec{a}$$

⟩ definitions

$$\frac{d}{dt}\int_V u_{\text{mech}}\, dV = -\frac{d}{dt}\int_V u_{\text{em}}\, dV - \oint_S \vec{S}\cdot d\vec{a}$$

⟩

$$\frac{d}{dt}\int_V \left(u_{\text{mech}} + u_{\text{em}}\right) dV = -\oint_S \vec{S}\cdot d\vec{a}$$

⟩ Gauss's theorem

$$\frac{d}{dt}\int_V \left(u_{\text{mech}} + u_{\text{em}}\right) dV = -\int_V \nabla\cdot \vec{S}\, dV$$

⟩

$$\frac{d}{dt}(u_{\text{mech}} + u_{\text{em}}) = -\nabla\cdot \vec{S}. \quad (5.38)$$

[25] The differential form of the continuity equation reads

$$\frac{\partial}{\partial t}\rho = -\nabla\cdot \vec{J}.$$

This equation is completely analogous to the continuity equation (Eq. 3.8).[25] Instead, of the charge density, we now have the energy density on the left-hand side. Moreover, instead of the current density $\vec{J}$, we now have the Poynting vector $\vec{S}$ on the right-hand side. While the current density describes how charges flow through our system, the Poynting vector describes how energy flows through the system. In the context of electromagnetic waves the Poynting vector $\vec{S}$ is especially important since it describes how energy gets transported by the electromagnetic field. Thus, to check the claim that electromagnetic waves transport energy, we have to calculate the Poynting vector corresponding to an explicit wave solution.

We again use our explicit solution discussed in Section 5.1:

$$\vec{S} \equiv \frac{1}{\mu_0}(\vec{E} \times \vec{B}) \qquad \text{Eq. 5.35}$$

↻ explicit solutions, Eq. 5.2, Eq. 5.8

$$= \frac{1}{\mu_0}\left(A\vec{e}_x \cos(kz - \omega t)\right) \times \left(\frac{k}{\omega}A\vec{e}_y \cos(kz - \omega t)\right)$$

$$= \frac{A^2 k}{\mu_0 \omega} \cos^2(kz - \omega t)\left(\vec{e}_x \times \vec{e}_y\right)$$

↻ $\vec{e}_x \times \vec{e}_y = \vec{e}_z$

$$= \frac{A^2 k}{\mu_0 \omega} \cos^2(kz - \omega t)\vec{e}_z$$

↻ $k^2/\omega^2 = \mu_0 \epsilon_0$, Eq. 5.29

$$= A^2 \sqrt{\frac{\epsilon_0}{\mu_0}} \cos^2(kz - \omega t)\vec{e}_z. \qquad (5.39)$$

This is clearly non-zero and therefore, as promised, electromagnetic waves do transport energy.

We can understand this result a little better by calculating the corresponding electromagnetic energy density (Eq. 5.37):[26]

[26] We will see in a moment why this is helpful.

$$u_{em} \equiv \frac{1}{2}\left(\epsilon_0 \vec{E}^2 + \frac{1}{\mu_0}\vec{B}^2\right)$$

↻ explicit solutions, Eq. 5.2, Eq. 5.8

$$= \frac{1}{2}\left(\epsilon_0\left(A\vec{e}_x \cos(kz - \omega t)\right)^2 + \frac{1}{\mu_0}\left(A\frac{k}{\omega}\vec{e}_x \cos(kz - \omega t)\right)^2\right)$$

↻

$$= \frac{1}{2}\left(\epsilon_0\left(A^2 \cos^2(kz - \omega t)\right) + \frac{1}{\mu_0}\left(A^2 \mu_0 \epsilon_0 \cos^2(kz - \omega t)\right)\right)$$

↻

$$= \epsilon_0 A^2 \cos^2(kz - \omega t). \qquad (5.40)$$

Therefore, we can write Eq. 5.39 as follows

$$\vec{S} = A^2 \sqrt{\frac{\epsilon_0}{\mu_0}} \cos^2(kz - \omega t) \vec{e}_z \quad \text{this is Eq. 5.39}$$

$$\circlearrowright \quad \frac{\epsilon_0}{\epsilon_0} = 1$$

$$= A^2 \sqrt{\frac{\epsilon_0}{\mu_0} \frac{\epsilon_0}{\epsilon_0}} \cos^2(kz - \omega t) \vec{e}_z$$

$$\circlearrowright \quad \text{Eq. 5.37}$$

$$= u_{\text{em}} \sqrt{\frac{1}{\epsilon_0 \mu_0}} \vec{e}_z$$

$$\circlearrowright \quad \sqrt{\frac{1}{\mu_0 \epsilon_0}} = c, \text{ Eq. 5.29}$$

$$= u_{\text{em}} c \vec{e}_z .$$

(5.41)

In words, this means that the energy flux $\vec{S}$ is simply the velocity of the wave $\vec{v} = c\vec{e}_z$ times the energy density u_{em}. This is completely analogous to how we can write the current density in terms of the charge density and the velocity.[27]

[27] $\vec{j} = \rho \vec{v}$, Eq. 2.12.

Before we move on, a few more comments on electromagnetic waves.

▷ There are lots of interesting phenomena associated with electrodynamical waves which are important, for example, in an engineering context. Using the explicit wave solutions discussed in the previous sections, it's possible to derive how electromagnetic waves scatter, get reflected and behave in matter.

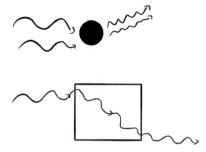

These kind of investigations lead to a huge sub-field of electrodynamics, commonly known as **optics**.

▷ We haven't talked about how electromagnetic waves can actually be produced. We know that a stationary electric charge produces a static electric field and an electric charge moving with *constant* velocity produces a static magnetic field. Electromagnetic waves are field configurations involving changing electric and magnetic field configurations.[28] The crucial question is therefore: How can we produce changing field configurations?

[28] Recall that the main mechanism behind electromagnetic waves is that a changing electric field produces a changing magnetic field and, in turn, a changing magnetic field produces a changing electric field.

The correct answer is: through accelerating charges. This means that an electromagnetic wave gets produced whenever we change the velocity of an electrically charged object. The charged object produces at each constant velocity a specific pattern in the magnetic field. Each time the velocity of the object changes, a new pattern is generated. Hence, by changing the velocity of a charged object, we can produce changing patterns in the magnetic field.

This is, for example, exactly what happens in an antenna. The current flowing through an antenna is alternating (AC), which means it periodically changes its direction. Therefore, the electrons in the antenna are constantly accelerated and decelerated when the direction of the current changes.

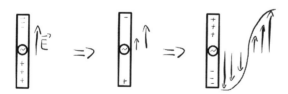

This is analogous to how we produce a wave in mechanics, for example, by shaking a rope:

Part III
Get an Understanding of Electrodynamics You Can Be Proud Of

"More recently, the principle of local gauge invariance has blossomed into a unifying theme that seems capable of embracing and even synthesizing all the elementary interactions."

Chris Quigg

PS: You can discuss the content of Part III with other readers and give feedback at www.nononsensebooks.com/edyn/part3.

In this final part, we will pick up a few loose ends:

▷ We will talk about the connection between Einstein's theory of special relativity and electrodynamics and how it helps us to understand the origin of Maxwell's equations.

▷ We will discuss gauge symmetry and how it can help to simplify calculations.

▷ In addition, we will investigate in what sense electrodynamics can be understood as a gauge theory. This is interesting since all fundamental interactions can be described using gauge theory. In other words, gauge theory allows us to understand all fundamental interactions from a common perspective.

▷ In the final chapter, we will talk about books which cover specific aspects in more detail.

In other words, if you're interested in fundamental physics, this part is where things get really interesting.

6

Special Relativity

In Section 5.5, we discovered a remarkable property of electromagnetic waves:[1]

> The speed of electromagnetic waves is always exactly $c = 1/\sqrt{\epsilon_0 \mu_0} = 2.9979 \times 10^8$ m/s.

While this little fact can easily go unnoticed, it leads to extremely wide-ranging consequences if taken seriously.[2] Most importantly, there are two noteworthy points here:

▷ The speed of electromagnetic waves is not infinite. In particular, this means that light needs some time to travel a given distance. Therefore, when we observe stars we actually see what they looked like many years ago since light needs many years to travel from distant stars to the earth.

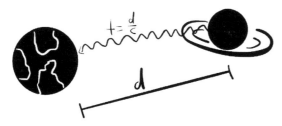

[1] We derived this in Eq. 5.29. However, take note that this is only true in a vacuum. An electromagnetic wave in matter travels with a smaller velocity since the constants ϵ_0 and μ_0 must be replaced with

$$\epsilon_0 \to \epsilon, \quad \mu_0 \to \mu.$$

The specific values of ϵ and μ, and therefore of the speed of electromagnetic waves, depend on the material at hand. Moreover, take note that the specific value 2.9979×10^8 m/s is not important since it depends on our choice of units. For example, if we measure length in feet, we have 9.836×10^8 feet/s. In theoretical physics it's often very convenient to use a special kind of units, known as natural units, where the speed of light is simply 1. The only thing that matters here is that c is nonzero and not infinite, which is true for all sensible systems of units.

[2] Historically, the importance of this property remained unnoticed until the famous Michelson-Morley experiment confirmed it experimentally. Afterward, Einstein was the first to take the result seriously and used it to develop his famous theory of special relativity.

In other words, when we observe the universe, we are actually observing the past and don't see what things are looking like right now. In addition, this means that we can only observe a finite part of the universe since some parts are so far away that light hasn't had enough time to reach us.

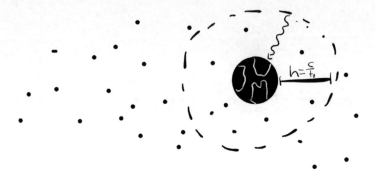

The age of the universe (the time since the Big Bang) is approximately $t_0 \approx 13.79$ billion years. This means that all parts of the universe further away than $d \approx ct_0$ are so far away that no light from there has ever reached the earth.

▷ The second important point is that we were able to determine the speed of light in full generality without making reference to anything. Usually the speed of an object depends on how we move relative to it, i.e. the frame of reference we are using. For example, imagine that an observer *standing* at a train station measures that a train moves with 50 $\frac{km}{h}$:

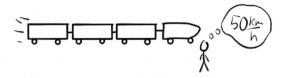

A second observer who runs with 15 $\frac{km}{h}$ parallel to the same train, measures that the train moves with 35 $\frac{km}{h}$.

Curiously, this does not happen for electromagnetic waves. Electromagnetic waves always travel with $c = 1/\sqrt{\epsilon_0 \mu_0} = 2.9979 \times 10^8$ m/s, no matter how you move.[3]

[3] As already mentioned above, the speed of light only has this value in free space and not if our wave moves in matter. The speed of electromagnetic waves in matter is lower.

The curious fact of nature that the speed of electromagnetic waves always has exactly the same value, leads to all kinds of strange consequences. Taking it seriously leads to Einstein's theory of **special relativity**. While there is no way we can discuss special relativity in detail, we should at least talk about the most famous phenomenon.

Let's imagine, a person sends a light pulse straight up where it is reflected by a mirror and finally reaches again the point from where it was sent:

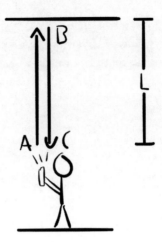

We record three important events:

> **A** : the light pulse leaves the starting point
> **B** : the light pulse is reflected by the mirror
> **C** : the light pulse returns to the starting point.

The time-interval between the two events **A** and **C** is[4]

$$\Delta t = t_C - t_A = \frac{2L}{c}, \qquad (6.1)$$

where L denotes the distance between the person and the mirror.

So far, nothing interesting has happened. But this changes quickly as soon as we consider how a second person observes exactly the same situation.

We imagine this second person moves with some constant speed u relative to the first person. For simplicity, we assume that the origins of the two coordinate systems coincide when the light pulse is send off (t_A). Moreover, we assume that each person stands at the origin of his coordinate system.

A first crucial observation is that the starting and end points of the light pulse have different coordinates for the second observer:

[4] Reminder: for a constant speed v we have $v = \frac{\Delta s}{\Delta t}$, where Δs is the distance and Δt the time interval. Therefore, we have $\Delta t = \frac{\Delta s}{v}$.

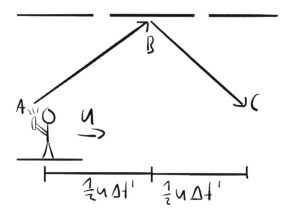

Mathematically, we have

$$x'_A = 0 \neq x'_C = u\Delta t' \quad \rightarrow \quad \Delta x' = u\Delta t', \quad (6.2)$$

where we use primed coordinates for the coordinate system associated with the second person. In words, this means that for this second person the light apparently also has moved in the x-direction. In contrast, for the first person

$$x_A = x_C \quad \rightarrow \quad \Delta x = 0. \quad (6.3)$$

Now, what's the time interval the second person measures between the event A and the event C?[5]

[5] It will become clear in a moment, why this is an interesting question.

As usual the time interval $\Delta t' = t'_C - t'_A$ can be calculated as the distance l divided by the speed of the light pulse c.

$$\Delta t' = \frac{l}{c} \quad (6.4)$$

The distance l is for this second observer no longer simply L, but we can calculate it using the Pythagorean theorem[6]

[6] See the triangle in the figure above.

$$l = 2\sqrt{\left(\frac{1}{2}u\Delta t'\right)^2 + L^2}. \quad (6.5)$$

The time interval measured by this second person is therefore

$$c\Delta t' = 2\sqrt{\left(\frac{1}{2}u\Delta t'\right)^2 + L^2} \qquad \text{this is Eq. 6.5 with Eq. 6.4 used on the LHS}$$

$$\Delta x' = u\Delta t',\ \text{Eq. 6.2}$$

$$= 2\sqrt{\left(\frac{1}{2}\Delta x'\right)^2 + L^2}$$

$$\Delta t' = \frac{2\sqrt{\left(\frac{1}{2}\Delta x'\right)^2 + L^2}}{c} \qquad (6.6)$$

[7] Reminder: $\Delta t = \frac{2L}{c}$, Eq. 6.1.

For $\Delta x' \neq 0$ this time interval is different from the time interval measured by the first person: $\Delta t' \neq \Delta t$.[7] In words, this means that two observers moving relative to each other do not agree on the time interval between the two events A and C!

This phenomenon is usually called **time-dilation**. Clocks tick differently for different observers and they count a different number of ticks between two events.

Analogously, it's possible to derive that different observers do not necessarily agree on the length of objects. This is known as **length contraction** and is another famous consequence of the constant speed of light. A third incredibly important consequence is that the speed of light c is an upper speed limit for everything physical. Unfortunately, as already mentioned above, discussing all these consequences in detail requires at least another book.[8]

[8] Some of the best books on special relativity are listed in Chapter 9.

6.1 The origin of Maxwell's equations

Historically, electrodynamics was developed before special relativity. And as discussed in the previous section, it's possible to derive special relativity using Maxwell's equations.

However, it's also possible to turn this line of thought upside down.

This means that we can use the experimental fact that the speed of light is constant for all observers to derive special relativity.[9] Then we can use special relativity to derive the correct equations of electrodynamics, i.e. Maxwell's equations.

[9] As mentioned above, historically this fact was indeed discovered experimentally by the Michelson-Morley experiment.

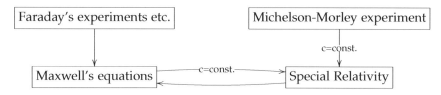

While the full story is far too long to include it here, let me sketch the main ideas.[10]

▷ Whenever we can perform a transformation of some object and the end result is indistinguishable, we are dealing with a symmetry.[11] We can then, motivated by the result of Michelson and Morley, search for all transformations which leave the speed of light invariant. These transformations are the symmetries of special relativity. Formally, this set of transformations is known as the **Poincaré group**.

[10] The full story can be found, for example, in

Jakob Schwichtenberg. *Physics from Symmetry*. Springer, Cham, Switzerland, 2018a. ISBN 978-3319666303

[11] A symmetry is a transformation which leaves the object in question unchanged.

▷ The second ingredient we need is the **Lagrangian formalism**. In modern physics, a theory is almost always defined in terms of a Lagrangian. Each theory (Newtonian mechanics, quantum mechanics, ...) corresponds to a specific Lagrangian. The correct equations describing the theory can then be derived by minimizing the so-called action, which is the integral over the Lagrangian.[12]

[12] In general, we search for extremal points of the action, i.e. sometimes the correct equations correspond to maxima of the action. The connection between Lagrangians and the corresponding equations of motion is given by the Euler-Lagrange equations.

[13] In general, only the action has to be invariant. However, if the Lagrangian is invariant, the action certainly is invariant too.

[14] This requirement is not really enough to derive the correct Lagrangian. Instead, the correct Lagrangian is the *simplest non-trivial* Lagrangian respecting the relevant symmetries.

[15] In mathematics this is known as representation theory.

[16] If we follow the same line of thought for a scalar field, we end up with the famous Klein-Gordon equation. For a third type of object our transformation can act on, known as spinors, we end up with the Dirac equation. Moreover, take note that there can be an additional term in the Lagrangian known as a mass term. Minimizing the action corresponding to the Lagrangian with this mass term leads to the so-called Proca equation. The absence of this mass term can be understood using gauge symmetry which is the topic of the next chapter.

[17] The equivalence of the two forms of the Lagrangian given here is demonstrated in Appendix A.17.1.

[18] The vacuum Maxwell equations are those without any electric current and without any charge density, i.e. the first two equations in Eq. 1.4 with $J=0$ and $\rho=0$.

▷ The main idea is that the Lagrangian must be invariant under all symmetries of the given theory.[13] This must be the case since we use the Lagrangian to derive the equations of motion. If the Lagrangian would change under a given symmetry transformation, we would get different equations of motions and hence the symmetry wouldn't be a symmetry. This observation is one of the main reasons why the Lagrangian formalism is so popular in modern physics.

▷ This allows us to derive the correct Lagrangian describing special relativity by demanding that it has to be invariant under all relevant symmetries.[14]

▷ In addition, by investigating the symmetries of special relativity carefully using the appropriate mathematical toolbox (group theory), we can also derive the correct Lagrangian describing electrodynamics. The crucial idea is that our symmetry transformation can act on different kinds of objects (scalars, vectors, etc.).[15] If we write down the simplest invariant Lagrangian involving a vector field A_μ, which describes nontrivial dynamics, we end up with the correct Lagrangian for electrodynamics:[16]

$$L_{\text{Maxwell}} = \frac{1}{2}(\partial^\mu A^\nu \partial_\mu A_\nu - \partial^\mu A^\nu \partial_\nu A_\mu) = \frac{1}{4}F^{\mu\nu}F_{\mu\nu}, \quad (6.7)$$

where $F_{\mu\nu}$ denotes the electromagnetic field tensor and $A_\mu = (\phi/c, A_i)^T$ denotes the electromagnetic potential.[17] Moreover, $A^\mu = (\phi/c, -A_i)^T$, $\partial_\mu = (\partial_0, \partial_i)^T$, and $\partial^\mu = (\partial_0, -\partial_i)^T$. By minimizing the corresponding action, we then end up with Maxwell's equations.

In fact, this way we only derive the inhomogeneous vacuum Maxwell equations:[18]

$$\nabla \cdot \vec{E} = 0$$

$$\nabla \times \vec{B} - \mu_0 \epsilon_0 \frac{\partial \vec{E}}{\partial t} = 0$$

The correct interplay between the electromagnetic potential A_μ and electric charges or currents can be derived using gauge symmetry. Formulated differently, using gauge symmetry, we can derive why and how the charge density and the current density show up on the right-hand side in the inhomogeneous Maxwell equations. We will discuss gauge symmetry and gauge theory in general in two following chapters. The homogeneous Maxwell equations

$$\partial_\lambda F_{\mu\nu} + \partial_\mu F_{\nu\lambda} + \partial_\nu F_{\lambda\mu} = 0 \qquad \text{(Eq. 1.1)} \qquad (6.8)$$

follow automatically from the definition of the electromagnetic field tensor $F_{\mu\nu}$ in terms of the potential A_μ:[19]

$$\partial_\lambda F_{\mu\nu} + \partial_\mu F_{\nu\lambda} + \partial_\nu F_{\lambda\mu}$$
$$= \partial_\lambda(\partial_\mu A_\nu - \partial_\nu A_\mu) + \partial_\mu(\partial_\nu A_\lambda - \partial_\lambda A_\nu) + \partial_\nu(\partial_\lambda A_\mu - \partial_\mu A_\lambda)$$
$$= \partial_\lambda\partial_\mu A_\nu - \partial_\lambda\partial_\nu A_\mu + \partial_\mu\partial_\nu A_\lambda - \partial_\mu\partial_\lambda A_\nu + \partial_\nu\partial_\lambda A_\mu - \partial_\nu\partial_\mu A_\lambda$$
$$= \partial_\lambda\partial_\mu A_\nu - \partial_\lambda\partial_\nu A_\mu + \partial_\mu\partial_\nu A_\lambda - \partial_\lambda\partial_\mu A_\nu + \partial_\lambda\partial_\nu A_\mu - \partial_\mu\partial_\nu A_\lambda$$
$$= \cancel{\partial_\lambda\partial_\mu A_\nu} - \cancel{\partial_\lambda\partial_\nu A_\mu} + \cancel{\partial_\mu\partial_\nu A_\lambda} - \cancel{\partial_\lambda\partial_\mu A_\nu} + \cancel{\partial_\lambda\partial_\nu A_\mu} - \cancel{\partial_\mu\partial_\nu A_\lambda}$$
$$= 0 \checkmark$$

$\circlearrowright$ $F_{\mu\nu} = \partial_\mu A_\nu - \partial_\nu A_\mu$

$\circlearrowright$

$\circlearrowright$ $\partial_\mu\partial_\nu = \partial_\nu\partial_\mu$

$\circlearrowright$

$\circlearrowright$

[19] It may seem strange that two of the four incredibly important Maxwell equations are, in some sense, so trivial. Maybe it helps to think about them as consistency conditions which supplement the inhomogeneous Maxwell equations. In addition, take note that there is a deep and beautiful geometrical reason for the homogeneous Maxwell equations. In mathematical terms, Eq. 6.8 is a Bianchi identity and in words, Bianchi identities always encode the geometrical fact that "the boundary of a boundary is zero". This is discussed in a bit more detail in Appendix A.16.1.

An important point is that in order to write down the correct Lagrangian which does not change under Poincaré transformations, we need to use A_μ (or equivalently $F_{\mu\nu} = \partial_\mu A_\nu - \partial_\nu A_\mu$, Eq. 2.22).

This is necessary because the electric field and the magnetic field are mixed through Poincaré transformations.

We can understand this by recalling that non-zero electric field strengths are created by stationary charges, while moving charges create non-zero magnetic field strengths. But a charge which is at rest for us, appears moving for a different observer who moves relative to the charge.

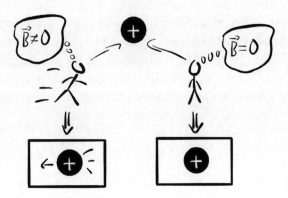

This means that we calculate that the charge produces no magnetic field strength since the charge is stationary, while the second observer finds a non-zero magnetic field strength.

Who is right?

We both are since the distinction between the electric and magnetic fields depends on how we observe the field. In reality, there is only one field called the electromagnetic field $F_{\mu\nu}$.[20] In practical terms, this means that to describe the electric and magnetic fields consistently, we need to use the electromagnetic tensor (Eq. 2.17)

$$F_{\mu\nu} = \begin{pmatrix} 0 & -E_1/c & -E_2/c & -E_3/c \\ E_1/c & 0 & -B_3 & B_2 \\ E_2/c & B_3 & 0 & -B_1 \\ E_3/c & -B_2 & B_1 & 0 \end{pmatrix}, \qquad (6.9)$$

which unifies them into a single object. The transformation which transforms our perspective into the perspective of the moving observer is a Poincaré transformation known as a **boost**.[21] What we have learned is therefore that a boost transforms an electric field strength into a magnetic field strength and vice versa. To write down a theory that is invariant under boost, we need to use $F_{\mu\nu}$ instead of the individual components $\vec{E}$ and $\vec{B}$.

Analogously, the electric potential ϕ and the vector potential A_i are mixed through Poincaré transformations. The four-vector

---

[20] Take note that this is analogous to how space and time are mixed by Poincaré transformations. The mixing of space and time is what we called time dilatation and space contraction previously. This mixing makes it necessary to introduce the four vector $x_\mu = (ct, \vec{x})^T$. In physical terms this means that we need to consider spacetime instead of space and time separately.

[21] In general, a boost is a transformation to a different constant velocity.

$A_\mu = (\phi/c, A_i)^T$ unifies these two potentials and allows us to write down an invariant Lagrangian (Eq. 6.7).

To summarize, we can say that from a relativistic perspective, the objects A_μ and $F_{\mu\nu}$ are better suited to describe electrodynamics than the electric and magnetic fields we usually talk about. In the following chapters, we will see how gauge symmetry allows us to understand electrodynamics and the objects A_μ and $F_{\mu\nu}$ from a completely new perspective.

7
Gauge Symmetry

In this chapter we will unclutter the notation a bit by setting $c = 1$. In other words, we don't write the speed of light everywhere. This is possible by using so-called "natural units". For our purposes it is sufficient to remember that if we want results in SI-units, we must add c in a few places. But such details aren't important for what follows because we are only interested in fundamental considerations.

In most introductory textbooks, gauge symmetry is either only mentioned in passing or described as a technical trick to make calculations simpler. However, from a modern perspective we can understand gauge symmetry as the fundamental defining feature of electrodynamics.

We will start by discussing gauge symmetry in somewhat abstract terms but afterwards discuss what it means and how it helps us to understand electrodynamics.

Gauge symmetry refers to the observation that we can't measure potentials like the electric potential A_0 or the magnetic potential $\vec{A}$ directly. We can see this, for example, by noting

[1] Alternatively, we can understand it by recalling that we can only measure potential energy differences. Potential energy differences remain unchanged by shifts of the potential

$\phi_1 - \phi_2 \to \phi_1 + \eta - (\phi_2 + \eta) = \phi_1 - \phi_2$,

where ϕ_1 and ϕ_2 denotes the potential energy of two objects in an electric field and we shifted the whole potential by a constant amount η.

[2] Reminder, Eq. 2.20:

$E_i = \partial_i A_0 - \partial_0 A_i$
$B_i = \epsilon_{ijk} \partial_j A_k$.

that the electric and magnetic field strengths remain completely unchanged if we shift our potentials by a constant amount:[1]

$$A_0 \to A_0 + \eta$$
$$A_i \to A_i + \xi_i. \qquad (7.1)$$

We can check this statement explicitly[2]

$$E_i = \partial_i A_0 - \partial_0 A_i \to \tilde{E}_i = \partial_i (A_0 + \eta) - \partial_0 (A_i + \xi_i)$$
$$= \partial_i A_0 + \underbrace{\partial_i \eta}_{=0} - \partial_0 A_i - \underbrace{\partial_0 \xi}_{=0}$$
$$= \partial_i A_0 - \partial_0 A_i$$
$$= E_i \checkmark \qquad (7.2)$$

where we used that ξ_i and η are constant. Analogously, we find

$$B_i = \epsilon_{ijk} \partial_j A_k \to \tilde{B}_i = \epsilon_{ijk} \partial_j (A_k + \xi_k)$$
$$= \epsilon_{ijk} \partial_j A_k + \epsilon_{ijk} \underbrace{\partial_j \xi_k}_{=0}$$
$$= \epsilon_{ijk} \partial_j A_k$$
$$= B_i \checkmark \qquad (7.3)$$

Upon closer inspection, we can discover that we have even more freedom. We can not only add constants to the potentials but also derivatives of an arbitrary scalar function $\eta(t, \vec{x})$. In particular, the electric and magnetic field strengths remain completely unaltered by the transformations

$$A_0 \to A_0 + \partial_0 \eta(t, \vec{x})$$
$$A_i \to A_i + \partial_i \eta(t, \vec{x}). \qquad (7.4)$$

Again, we can check this explicitly:

$$E_i = \partial_i A_0 - \partial_0 A_i \to \tilde{E}_i = \partial_i (A_0 + \partial_0 \eta(t, \vec{x})) - \partial_0 (A_i + \partial_i \eta(t, \vec{x}))$$
$$= \partial_i A_0 + \partial_i \partial_0 \eta(t, \vec{x}) - \partial_0 A_i - \partial_0 \partial_i \eta(t, \vec{x})$$
$$= \partial_i A_0 + \cancel{\partial_0 \partial_i \eta(t, \vec{x})} - \partial_0 A_i - \cancel{\partial_0 \partial_i \eta(t, \vec{x})}$$
$$= \partial_i A_0 - \partial_0 A_i$$
$$= E_i \checkmark \qquad (7.5)$$

and

$$\begin{aligned}
B_i = \epsilon_{ijk}\partial_j A_k \to \tilde{B}_i &= \epsilon_{ijk}\partial_j(A_k + \partial_k \eta(t,\vec{x})) \\
&= \epsilon_{ijk}\partial_j A_k + \underbrace{\epsilon_{ijk}\partial_j \partial_k \eta(t,\vec{x})}_{=0} \\
&= \epsilon_{ijk}\partial_j A_k \\
&= B_i \;\checkmark
\end{aligned} \qquad (7.6)$$

where we used that ϵ_{ijk} is antisymmetric but $\partial_j \partial_k$ is symmetric under the switching of the indices $j \leftrightarrow k$.[3] Alternatively we can also see this by writing the equation as a vector equation

$$\begin{aligned}
\vec{B} = \vec{\nabla}\times\vec{A} \to \vec{\nabla}\times(\vec{A} + \vec{\nabla}\eta(t,\vec{x})) &= \vec{\nabla}\times\vec{A} + \underbrace{\vec{\nabla}\times\vec{\nabla}\eta(t,\vec{x})}_{=0} \\
&= \vec{\nabla}\times\vec{A} \\
&= \vec{B} \;\checkmark
\end{aligned} \qquad (7.7)$$

where we used that the curl of a gradient is always zero.[4]

The fact that everything we can measure remains completely unchanged by the transformations in Eq. 7.4 is known as **gauge freedom** or **gauge symmetry**.[5]

Gauge symmetry is an extremely useful discovery for the following reason. We can describe electrodynamical systems using the potentials A_0, A_i. However, the physics in our system remains completely unchanged by a **gauge transformation** (Eq. 7.4). This means that we can use gauge transformations to simplify our calculations![6]

For example, it is always possible to choose a scalar function $\eta(t,\vec{x})$ in such a way that $\vec{\nabla}\cdot\vec{A} = 0$. This means that if we make this particular choice for our scalar function $\eta(t,\vec{x})$, all terms involving $\vec{\nabla}\cdot\vec{A}$ immediately drop out from all of our equations and this often simplifies our calculations tremendously. Choosing a specific scalar function is commonly called **choosing a gauge**. The choice leading to $\vec{\nabla}\cdot\vec{A} = 0$ is known as the **Coulomb gauge**. This choice is useful, for example, as already

---

[3] This is an important general result. Every time we have a sum over something symmetric in its indices multiplied by something antisymmetric in the same indices, the result is zero:

$$\sum_{ij} a_{ij} b_{ij} = 0$$

if $a_{ij} = -a_{ji}$ and $b_{ij} = b_{ji}$ holds for all i,j. We can see this by writing

$$\sum_{ij} a_{ij} b_{ij} = \frac{1}{2}\left(\sum_{ij} a_{ij} b_{ij} + \sum_{ij} a_{ij} b_{ij}\right)$$

We are free to rename our indices $i \to j$ and $j \to i$, which we use in the second term

$$\to \sum_{ij} a_{ij} b_{ij} = \frac{1}{2}\left(\sum_{ij} a_{ij} b_{ij} + \sum_{ij} a_{ji} b_{ji}\right)$$

Then we use the symmetry of b_{ij} and antisymmetry of a_{ij}, to switch the indices in the second term, which yields

$$\to \sum_{ij} a_{ij} b_{ij} = \frac{1}{2}\left(\sum_{ij} a_{ij} b_{ij} + \sum_{ij} \underbrace{a_{ji}}_{=-a_{ij}} \underbrace{b_{ji}}_{=b_{ij}}\right)$$

$$= \frac{1}{2}\left(\sum_{ij} a_{ij} b_{ij} - \sum_{ij} a_{ij} b_{ij}\right) = 0$$

[4] This is discussed in detail in Appendix A.16.

[5] As mentioned above, a symmetry always refers to a set of transformations which leaves our system unchanged.

[6] Don't worry if this seems like a cheap trick or otherwise strange to you because we will discuss what is really going on in the following chapter.

discussed in Section 4.2.3, because it simplifies the Poisson equation for $\vec{A}$.

To understand how this works, imagine that we have found a potential $\vec{A}$ which yields the correct magnetic field describing our system $\vec{B} = \nabla \times \vec{A}$. In general, the divergence of this potential $\vec{A}$ is some non-vanishing scalar function $\nabla \cdot \vec{A}(\vec{x}) = f(\vec{x})$. Now, we can use our gauge freedom (Eq. 7.4) to modify the potential

$$\vec{A}' = \vec{A} + \nabla \eta(t, \vec{x}). \tag{7.8}$$

The divergence of this new potential reads

$$\nabla \cdot \vec{A}'(t, \vec{x}) = \nabla \cdot \vec{A}(t, \vec{x}) + \nabla \cdot \nabla \eta(t, \vec{x})$$
$$= f(t, \vec{x}) + \nabla \cdot \nabla \eta(t, \vec{x}). \tag{7.9}$$

Therefore, we can indeed achieve that $\nabla \cdot \vec{A}'(t, \vec{x}) = 0$ by using a gauge transformation (Eq. 7.4) with a scalar function $\eta_{\text{sol}}(t, \vec{x})$ which we can find by solving

$$\nabla \cdot \nabla \eta_{\text{sol}}(t, \vec{x}) = -f(t, \vec{x}) \tag{7.10}$$

since

$$\nabla \cdot \vec{A}'(t, \vec{x}) \stackrel{!}{=} 0 \quad \Rightarrow \quad f(t, \vec{x}) + \nabla \cdot \nabla \eta(t, \vec{x}) \stackrel{!}{=} 0. \tag{7.11}$$

It can be shown rigorously that a solution to Eq. 7.10 always exists. In practice, we don't have to find the function $\eta_{\text{sol}}(t, \vec{x})$ explicitly, but instead can simply use the knowledge that a function with the desired properties exists and set $\nabla \cdot \vec{A}(t, \vec{x}) = 0$.

So far, gauge symmetry may seem like a boring mathematical trick, like voodoo or even both. However, gauge symmetry is really much more than that. Gauge symmetry is one of the guiding themes in modern physics and allows us to understand all fundamental interactions from a common perspective.[7]

[7] Be warned that there are lots of discussions about in what sense gravity can be understood using gauge symmetry, but a proper discussion of this question is far beyond the scope of this book.

While, once more, the full story is far too long to include it here, we can get at least some idea of how gauge symmetry and fundamental interactions are connected by considering a simple toy model.[8] We will talk about this in the next chapter. However before we discuss it, we need to talk about a second puzzle piece which is necessary to understand gauge symmetry: quantum mechanics.[9]

Of course, we can't discuss quantum mechanics properly here.[10] But luckily we only need one small fact to move forward: in quantum mechanics, we describe what is going on using a *complex* function $\Psi(\vec{x})$ called the **wave function**. However, everything we can measure is a *real* number. This means that the relationship between quantities we can measure O and the wave function Ψ is always of the form $O \simeq \Psi^\star \Psi$, where the superscript $\star$ denotes complex conjugation. A direct consequence of this observation is that we have some freedom in our wave functions, analogous to how we have some freedom in our electric and magnetic potentials. In particular, we always have the freedom to multiply our wave functions by a **phase factor**

$$\Psi \to \tilde{\Psi} = e^{i\varphi}\Psi, \qquad (7.12)$$

since

$$O \simeq \Psi^\star \Psi \quad \Rightarrow \quad \tilde{\Psi}^\star \tilde{\Psi} = (e^{i\varphi}\Psi)^\star e^{i\varphi}\Psi$$
$$= e^{-i\varphi}\Psi^\star e^{i\varphi}\Psi \qquad (e^{ix})^\star = e^{-ix}$$
$$= \Psi^\star \underbrace{e^{-i\varphi}e^{i\varphi}}_{=1} \Psi$$
$$= \Psi^\star \Psi = O \quad \checkmark \qquad (7.13)$$

One of the deepest observations in modern physics is that there is *direct* connection between the freedom in our wave function (Eq. 7.12) and the freedom in our electromagnetic potential (Eq. 7.4). How this comes about is the topic of the next chapter.

[8] A complete discussion can be found in any textbook on gauge theory and, for example, in

Jakob Schwichtenberg. *Physics from Symmetry*. Springer, Cham, Switzerland, 2018a. ISBN 978-3319666303

[9] This already demonstrates how gauge symmetry ties together seemingly unrelated parts of physics.

[10] If you're interested in quantum mechanics, you might enjoy reading

Jakob Schwichtenberg. *No-Nonsense Quantum Mechanics*. No-Nonsense Books, Karlsruhe, Germany, 2018b. ISBN 978-1719838719

8

Electrodynamics as a Gauge Theory

In this chapter, we will discuss the interplay between gauge symmetry and fundamental interactions in a bit more detail. Since we can't discuss quantum mechanics and gauge theory in full detail, we will use a relatively simple toy model.[1]

At first, it may seem like this toy model doesn't really have anything to do with quantum mechanics, electrodynamics or even physics in general. However, this couldn't be further from the truth. The interplay between interactions and gauge symmetry in the toy model is *completely analogous* to how things work in quantum mechanics and electrodynamics.

We will discuss the connection between the toy model and our physical theories explicitly at the end of the chapter. For the moment, just take note that it's not accidental that we use the same symbols $A_\mu, F_{\mu\nu}, J_\mu$ to describe the toy model.

Before we dive in, a short warning: we will discuss quite a bit of quantum mechanics in the following sections. But don't worry if you don't yet know anything about quantum mechanics and

[1] You can find a much more complete discussion of the ideas outlined in this chapter in

Jakob Schwichtenberg. *Physics from Finance*. No-Nonsense Books, Karlsruhe, Germany, 2019. ISBN 978-1795882415

can't follow all of the arguments. Just keep reading and glimpse over the parts that you don't understand. The most important ideas will still make sense and you can reread everything you didn't understand once you know more about quantum mechanics.

8.1 Symmetries intuitively

Before we can discuss gauge symmetry, we have to talk about symmetries in general.

So first of all, what is a symmetry?

Imagine a friend stands in front of you and holds a perfectly round ball in her hand. Then you close your eyes, your friend performs a transformation of the ball and afterward you open your eyes again. If now, for example, she rotates the ball while your eyes are closed, it is impossible for you to find out whether she did anything at all. Hence, rotations are symmetries of the ball.

In contrast, if she holds a cube, only very special rotations can be done without you noticing it. In general, all transformations which, in principle, change something but lead to an indistinguishable result are symmetries. Formulated differently, a symmetry takes us from one state to a different one which happens to have the same properties.[2]

[2] In contrast, a redundancy takes us from one description of a state to another description of the same state. This will be discussed in detail below.

It's important to take note that with this definition, symmetries are *observable* properties of objects or systems. This is especially important in the context of gauge symmetries since here we need to be careful what transformations are symmetries and which are mere redundancies.

This point can be confusing at first and to understand it, we need to talk about subsystems and global or local transformations of them.

8.1.1 Global vs. local symmetries

A **global transformation** is one in which the whole ship is, for example, rotated as opposed to a transform in which only a part of it is rotated (which would be a **local transformation**).

Clearly a global rotation of a ship takes us from one state to a physically different one since we are actively rotating the ship. However, if there is a physicist inside the ship with no possibility to look outside, there is no way for him to find out whether he is in the original or the rotated ship.[3]

[3] This thought experiment is known as Galileo's ship experiment.

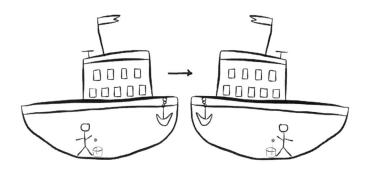

Therefore, the subsystem has a global rotational symmetry. The crucial point is that, in principle, it would be possible to detect a difference since the ship and the rotated ship are two distinct states.

In contrast, there is no way that how we could ever detect a global rotation of the whole universe. This is why the notion of subsystems is essential in this context [4]. Moreover, in physics

[4] D. Wallace and Hilary Greaves. Empirical consequences of symmetries. *British Journal for the Philosophy of Science*, 65(1):59–89, 2014

we always consider sufficiently isolated subsystems, even if this is not explicitly stated and mathematically we take the limit $|x| \to \infty$.

Now what about local transformations?

Our physicist inside the ship would immediately notice if we rotate only a part of the ship, e.g., a specific apparatus.

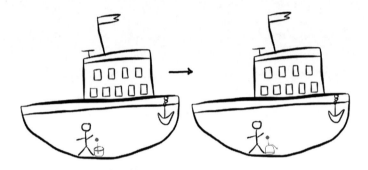

Therefore, we don't have local rotational symmetry. This is an important observation in the context of gauge theories where often special emphasis is put on local transformations. While global symmetries are common, local symmetries are rarely observed in nature.[5]

[5] An example of a system with a local symmetry is a grid of completely black balls. We can rotate each ball individually and it would be impossible to tell the difference.

Now, after these preliminary remarks we can start to discuss gauge symmetries. In the following section, we introduce all relevant notions in the context of a simple toy model. This allows us to understand gauge symmetries without confusing abstractions and complicated formulas.

8.2 A toy gauge theory

The toy model we will use in the following describes a simplified financial market. It consists of several countries and the basic process we try to describe is that money can be carried around. This setup is arguably the simplest setup where a gauge symmetry shows up.[6]

[6] The connection between financial markets and gauge symmetries was first put forward in [Ilinski, 1997] and later popularized in [Young, 1999] and [Maldacena, 2016].

First of all, let's imagine that we have a common currency for several countries. For concreteness, we call this currency euro and the countries Germany, France, the United Kingdom and Italy.

In addition, we consider this subsystem of the whole world isolated and assume that there is a trader who only does business within these countries. A crucial observation is that this subsystem has a **global gauge symmetry** since the absolute value of fiat money is, in general, not determined. We can change the units of the currency globally without any physical effect. For example, absolute prices are not fixed by anything, only relative prices are.[7]

[7] At least in our toy model, changing the value of the currency has no effect since we imagine that all prices and wages are simply adjusted automatically. Of course, in the real world there could be psychological effects since people get used to certain prices.

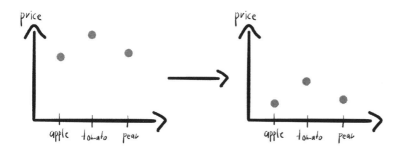

Imagine for example that our trader sells three tomatoes at €1 each and then uses this money to buy six apples at €0.5 each.

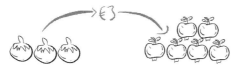

The end result of such a process is completely unchanged if, for example, the government decides to print lots of euros such that the value of each euro drops by a factor of ten. Afterwards, the trader gets €10 for each tomato but then needs to pay €5 per apple. So again, our trader starts with three tomatoes and ends up with six apples.

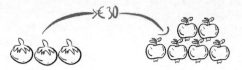

This is no longer true if only one country, say France, decides to print lots of euros and give it to its people. Such a local change of the value of our currency certainly has a noticeable effect. The prices would only increase in France and our trader could sell his tomatoes in France at €10 each and buy apples at €0.05 in Germany.[8] So when our trader starts again with three tomatoes, he will now end up with sixty apples. In technical terms this example shows that our subsystem is not invariant under local transformations of the value of our currency. Alternatively, we can say that we have a global symmetry but not a local one.

[8] Take note that so far our toy model is completely static and therefore there is no dynamically adjusting of the prices.

Now, in the real world there is more than one currency and we can imagine that we can change the numbers on the local currencies arbitrarily. Before we can discuss what this means for the symmetries of our system, we need to talk about the difference between two different kinds of transformations.[9]

[9] To spoil the surprise: only invariance under active transformations means that we are dealing with a symmetry. Invariance under passive transformations merely indicates a redundancy in our description.

8.2.1 Active vs. passive transformations

So far, we only considered active transformations. We discussed what happens when we rotate Galileo's ship or when a country actually prints new money. These are real physical transformations.

However, there is also a different kind of transformation called passive transformations. A passive transformation is a change in how we *describe* a given system.

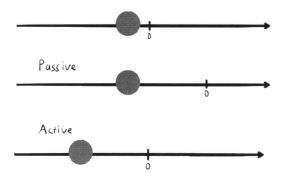

For example, we can describe Galileo's ship using curvilinear coordinates or a rotated coordinate system. Of course, such transformations never have any physical effect.[10] In our financial toy model, a passive transformation is a change of the money coordinate system, for example, when we drop digits from the euro.

To summarize: passive transformations relate different descriptions of the same physical situation, while active transformations relate different physical situations.

A lot of confusion surrounding gauge symmetries can be traced back to confusion about these two kinds of transformations and it's crucial to keep them separate.

With this in mind, we are finally ready to discuss another incredibly important distinction which is necessary to make sense of gauge transformations.

[10] If we write down the equations describing our system appropriately this becomes immediately clear. We will discuss this explicitly below.

8.2.2 Symmetries vs. redundancies

As mentioned in the last section, passive transformations are always possible without physical implications since they are merely a change in our description of the system.

However, if we now reconsider our financial toy model, it is not immediately clear how this comes about. So far, locally adding or dropping zeroes from a given currency clearly makes a difference. But this is only a problem in our description and not a physical effect. When we locally change the "money coordinate system" in a given country, we effectively introduce a new currency. This new currency then has a specific value relative to the original currency.

For concreteness, we now introduce independent local currencies in, say Germany, which we call Deutsche Mark (DM). Moreover, we introduce Francs (F) in France, in England Pounds (P) and Lira (L) in Italy.

This is only possible if we introduce some sort of bookkeepers who keep track of the values of local currencies and are able to exchange one currency for another. If we assume that the bookkeepers always adjust their exchange rates perfectly whenever the value of our local currency changes, such changes have no noticeable effect.

For example, let's assume that the exchange rates are

$$\begin{aligned} DM/P &= 1 \\ P/F &= 2 \\ F/L &= 10 \\ DM/L &= 20 \, . \end{aligned} \qquad (8.1)$$

Now, if a trader starts with $1DM$, he can trade it for $1P$, then use it to trade it for $2F$, then trade these for $20L$ and finally trade these back for $1DM$.

When a local currency changes, the bookkeepers simply adjust their exchange rates accordingly and the situation remains the same. Thanks to these bookkeepers, we can see that such passive transformations really make no difference as it should be.

What we did here is a typical example of what is necessary to make the invariance under local passive transformations manifest. When we want to describe a system in such a way that arbitrary descriptions are allowed, we need to add new (mathematical) ingredients. A passive transformation, by definition, cannot lead to a new physical situation and our bookkeepers make sure that this is indeed the case.[11]

An important point is that, so far, our bookkeepers aren't dynamical actors which start actions on their own. In addition, they do not have any physical influence on the dynamics of the system. Instead their only task is to keep track of the local coordinate changes.

Since it is important to keep passive and active transformations separate, we also need to keep invariance under them separate. We call invariance under passive transformations a **redundancy** and invariance under active transformations a **symmetry**.

Using this definition, we can say that after the introduction of the bookkeepers *our description* of the system has a local redun-

[11] As we will discuss below, the mathematical tools that allow us to describe a given system using arbitrary coordinate system are called connections. In physics, we call these connections gauge potentials.

dancy.

Now what about local symmetry, i.e. invariance under *active* local transformations?[12]

[12] As discussed above, without the bookkeepers our financial toy system was not invariant under active local transformation.

At first glance it may seem as if after the introduction of the bookkeepers we also have a local symmetry. However, this is not the case. The bookkeepers do not react to active transformations. So far, they are purely mathematical objects which only exist to keep track of local passive transformations. An active transformation is a real physical change and our purely mathematical bookkeepers cannot induce a physical change on their own which would cancel the active transformation and make the system invariant.[13]

[13] As we will discuss below, it is possible to turn the bookkeepers into active agents which are capable of inducing physical changes. Then we have to discuss the question of local symmetry again. However, it is already clear at this point that the answer crucially depends on the rules according to which the bookkeepers behave. In physics, we have to derive these rules from experiments.

At this point we can already give a preliminary definition of the notion of local gauge symmetry, which, however, will be refined later.

A **local gauge symmetry** is not really a symmetry but rather a redundancy which appears in our description. The redundancy that exists in our description of the financial toy example after the introduction of the bookkeepers is an example of a local gauge symmetry.

8.3 Gauge dynamics

So far, our bookkeepers are purely mathematical ingredients which we introduced to make our description invariant under local passive transformations.

Now, if we go back to our financial toy example, we can imagine that bookkeepers can influence the dynamics of a system and even become dynamical actors on their own.

First of all, we can imagine that there are imperfections in the exchange rates. If this is the case, we can imagine that our trader no longer trades goods, but starts to trade money. Now that we have local currencies and possibly imperfections in the exchange rates, this can be a lucrative business. This is an explicit example of how our bookkeepers can influence the dynamics of the system.

For example, let's imagine the exchange rates set by the bank are as follows

$$DM/P = 1$$
$$P/F = 2$$
$$F/L = 10$$
$$DM/L = 10. \tag{8.2}$$

Now our trader is able to earn money simply by exchanging money. If he starts with $1DM$, he can trade it for $1P$, then use the pound to trade it for $2F$, then trade these for $20L$ and finally trade these for $2DM$.

In the financial world this is known as an **arbitrage opportunity**.[14]

But wait, the exchange rates depend on the values of the local currencies which we can change at will ... Does this mean that the amount of money our trader earns depends on these arbitrary choices?

First of all, our exchange rates indeed do change when, for example, Italy decides to drop a zero from their currency. We

[14] Arbitrage is a risk-free possibility to earn money. In mathematics, we call the quantity which we here use to describe arbitrage opportunities the **curvature**.

need to consistently take such a change of the local money coordinate system $L \to \tilde{L} = L/10$ into account and this means that we need to adjust the exchange rates accordingly:

$$DM/P = 1$$
$$P/F = 2$$
$$F/L = 10 \quad \to \quad F/\tilde{L} = F/(L/10) = 1$$
$$DM/L = 10 \quad \to \quad DM/\tilde{L} = DM/(L/10) = 1. \qquad (8.3)$$

However, the amount of money our trader earns is unchanged by such a re-scaling!

If he starts again with $1DM$, he can still trade it for $1P$, which he can trade for $2F$, then trade these for $2L$ and finally trade these for $2DM$. The final result is the same as before.

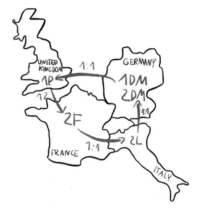

What we have learned here is a crucial aspect of every **gauge theory**. An important task is to find quantities which do not depend on local conventions. For example, the possibility to earn risk-free money is independent of how we choose our local coordinate system.

There is another way to make risk-free money which involves interest rates. Let's assume that the interest rate is 2 % per year for the Deutsche Mark. Moreover, we assume that the exchange rate changes over time and is now

$$DM/L = 10 \qquad (8.4)$$

and in one year
$$DM/L = 5. \qquad (8.5)$$

In this scenario it's possible to earn money as follows.

Our trader can borrow $1DM$ and after 1 year he has to pay $1.02DM$ back. Moreover, he can now exchange his $1DM$ for $10L$. After one year he can then exchange his $10L$ for $2DM$. Therefore, after paying back the $1.02DM$ he has made $0.98DM$ in profit.

This is another example of a risk-free arbitrage trade.

The crucial point in these examples is that the situation in our toy model is a very different one whether there is an arbitrage opportunity or not. Formulated differently, when there is an arbitrage opportunity, our bookkeepers are more than mere mathematical tools since their exchange rates actively shape the dynamics of the system.[15] However, so far, our bookkeepers are still not dynamical. Their exchange rates don't change and while the bookkeepers have a real effect on the system they only yield the static background in front of which all dynamical actors do their business.

However, next we can imagine that our bookkeepers become dynamical actors. In the real world, banks calculate exchange rates, but they are also institutions that live in the real world, not only in our description of financial markets, and make decisions dynamically.

Promoting bookkeepers to dynamical objects which follow their own rules is a crucial step to make our toy model more realistic.[16]

While we can always introduce local redundancies into our description of a given system, we are only dealing with a **gauge theory** when the bookkeepers are dynamical objects in the system and not only artifacts of our description. The defining feature of a gauge theory is that the bookkeepers, which are necessary to make the description of the system invariant under

[15] In physics, the situation with an arbitrage opportunity corresponds to a system with non-zero curvature. But take note that this curvature is not necessarily dynamical. Only when the curvature is dynamical are we dealing with a gauge field. Otherwise we are describing physics happening in a static curved space.

[16] In physics, a dynamical object is something with its own equation of motion.

local passive transformations, are physical actors that shape the dynamics of the system. In practice this means that exchange rates are adjusted dynamically depending on what else happens in the system. We will discuss the explicit rules according to which this happens in Section 8.3.2.

Before we move on and discuss gauge symmetries and gauge theories in more mathematical terms, there is a question which we still need to answer.

Now that we have promoted our bookkeepers to dynamical agents, do we end up with a local symmetry?

Our bookkeepers are now parts of the system and not just of our description and they can therefore induce real changes. Thus, in principle, it's possible that the action of an active transformation is canceled through a bookkeeper. However, this is not automatically the case and in general, we still don't have a local symmetry. This will become a lot clearer as soon as we discuss gauge symmetries in the context of physical theories like electrodynamics.

But first we have to introduce the proper mathematical notions which can all be easily understood in the context of our toy model.

8.3.1 Mathematical description of the toy model

Mathematically, we imagine that our **countries** live on a lattice. Each point on the lattice is labelled by d-numbers: $\vec{n} = (n_1, n_2, \cdots, n_d)$. In other words, each country can be identified by a vector $\vec{n}$ which points to its location.

We can move from one country to a neighboring country by using a **basis vector** $\vec{e}_i$, where i denotes the direction we are moving. For example, $\vec{e}_2 = (0, 1, 0 \cdots, 0)$.

We denote the **exchange rates** between the country labelled by the vector $\vec{n}$ and its neighbor in the i-direction by $R_{\vec{n},i}$. For example, if the country at the location labelled by $\vec{n}$ uses Deutsche Marks and its neighbor in the 2-direction uses Francs, $R_{\vec{n},2}$ tells us how many Francs we get for each Deutsche Mark.

In physics, we usually introduce the corresponding logarithms

$$R_{\vec{n},i} \equiv e^{A_i(\vec{n})}, \tag{8.6}$$

where $A_i(\vec{n}) \equiv \ln(R_{\vec{n},i})$.

The next ingredient that we need is a notation for gauge transformations. In our toy model, a gauge transformation is a change of currencies and directly impacts, for example, the exchange rates. We use the notation $f(\vec{n})$ to denote a change of the currency in the country at $\vec{n}$ by a factor of $f(\vec{n})$. In addition, we again introduce the corresponding logarithm

$$f(\vec{n}) \equiv e^{\epsilon(\vec{n})}. \tag{8.7}$$

In general, when we perform such a gauge transformation in the country labelled by $\vec{n}$ and also in the neighboring country in the i-direction, the corresponding exchange rate changes as follows[17]

$$R_{\vec{n},i} \to \frac{f(\vec{n}+\vec{e}_i)}{f(\vec{n})} R_{\vec{n},i}. \tag{8.8}$$

In terms of the logarithms this equation reads

$$R_{\vec{n},i} = e^{A_i(\vec{n})} \to \frac{f(\vec{n}+\vec{e}_i)}{f(\vec{n})} R_{\vec{n},i}$$

↓ Eq. 8.7

$$= \frac{e^{\epsilon(\vec{n}+\vec{e}_i)}}{e^{\epsilon(\vec{n})}} e^{A_i(\vec{n})}$$

↓

$$= e^{A_i(\vec{n})+\epsilon(\vec{n}+\vec{e}_i)-\epsilon(\vec{n})} \tag{8.9}$$

and we can conclude

$$A_i(\vec{n}) \to A_i(\vec{n}) + \epsilon(\vec{n}+\vec{e}_i) - \epsilon(\vec{n}). \tag{8.10}$$

We learned above that an import aspect of the system is whether arbitrage opportunities exist. An arbitrage opportunity exists

[17] The currency in the country at $\vec{n}$ gets multiplied by $f_{\vec{n}}$ and the currency in its neighbor in the i-direction by $f_{\vec{n}+\vec{e}_i}$. Therefore, the exchange rate gets modified by the ratio of these two factors.

when we can trade currencies in such a way that we end up with more money than we started with. But we can only make such a statement when the starting currency and the final currency are the same. Only then can we be certain whether the final amount of money is larger than the initial amount. Therefore, we need to trade money in a loop.

The total gain we can earn by following a specific loop can be quantified by

$$G = R_{\vec{n},i} R_{\vec{n}+\vec{e}_i,j} \frac{1}{R_{\vec{n}+\vec{e}_j,i}} \frac{1}{R_{\vec{n},j}}. \tag{8.11}$$

When this **gain factor** is larger than one, we can earn money by trading money following the loop, if it is smaller than one we lose money.

To understand the definition of the gain factor, imagine that we start with $1DM$. We trade it for Pounds and $R_{\vec{n},1} = 1$ tells us that we get $1P$. Afterwards, we trade our Pounds for Francs and $R_{\vec{n}+\vec{e}_1,2} = 2$ tells us that we get in total $2F$. Afterwards, we trade our Franc for Lira. $R_{\vec{n}+\vec{e}_2,1} = 10$ tells us that we get $1/10$ Franc for each Lira. Hence, we have to calculate $2F/R_{\vec{n}+\vec{e}_2,1} = 20L$. Finally, we use that $R_{\vec{n},2} = 10$ tells us that we get $10L$ for each Deutsche Mark and therefore calculate $20L/R_{\vec{n},2} = 2DM$.[18]

[18] The crucial point is that our exchange rates $R_{\vec{n},i}$ always tell us how many of the currency at the neighboring country in the i-direction we get for each unit of the local currency at the country at $\vec{n}$. Hence, we sometimes have to divide by the corresponding exchange rate to calculate the resulting amount of a new currency.

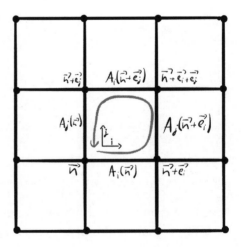

Once more we introduce the corresponding logarithm

$$G \equiv e^{F_{ij}(\vec{n})} \qquad (8.12)$$

and again, we can rewrite our equation in terms of the logarithms[19]

$$F_{ij}(\vec{n}) = A_j(\vec{n} + \vec{e}_i) - A_j(\vec{n}) - [A_i(\vec{n} + \vec{e}_j) - A_i(\vec{n})]. \qquad (8.13)$$

[19] In physics F_{ij} is directly related to components of the **magnetic field**. For example, $F_{12} = -B_3$.

A crucial consistency check is that G and F_{ij} are unchanged by gauge transformations. We already argued above that an arbitrage opportunity is something real and thus cannot depend on local choices of the coordinate system. Quantities like this are usually called **gauge invariant**. So in words, G and $F_{ij}(\vec{n})$ encode what is physical in the structure of exchange rates.[20] Moreover, an important technical observation is that $F_{ij}(\vec{n})$ is antisymmetric: $F_{ij}(\vec{n}) = -F_{ji}(\vec{n})$, which follows directly from the definition.

[20] A single exchange rate $A_i(\vec{n})$ is gauge dependent and can therefore, for example, be set to zero simply by changing a local money coordinate system.

So far, we only talked about spatial exchange rates. However, there are also temporal exchange rates, i.e. interest rates. A clever trick to incorporate this is to introduce time as the zeroth-coordinate like we do in special relativity. In other words, in addition to specific locations (countries) our lattice now contains copies of these locations at different points in time. This means that a point on the lattice is specified by $d + 1$ coordinates $\vec{n} = (n_0, n_1, n_2, \cdots, n_d)$ and the zeroth component indicates the point in time.

Then, Eq. 8.13 reads[21]

$$F_{\mu\nu}(\vec{n}) = A_\nu(\vec{n} + \vec{e}_\mu) - A_\nu(\vec{n}) - [A_\mu(\vec{n} + \vec{e}_\nu) - A_\mu(\vec{n})], \qquad (8.14)$$

where previously $i, j \in \{1, 2, \ldots, d\}$ and now $\mu, \nu \in \{0, 1, 2, \ldots, d\}$.

[21] The non-vanishing components of $F_{\mu\nu}(\vec{n})$ with either $\mu = 0$ or $\nu = 0$ are directly related to what we call electric field in physics. For example, $F_{10} = E_1$.

In the continuum limit, where the lattice spacing goes to zero, Eq. 8.10 becomes[22]

$$A_\mu(x_\mu) \to A_\mu(x_\mu) + \frac{\partial \epsilon}{\partial x^\mu} \qquad (8.15)$$

and Eq. 8.14 reads[23]

[22] To understand this, take note that in Eq. 8.14 we get in this limit the difference quotient. Compare this equation with Eq. 7.4.

[23] This is exactly how we defined the field strength tensor in Section 2.4.

$$F_{\mu\nu}(x_\mu) \equiv \frac{\partial A_\nu}{\partial x^\mu} - \frac{\partial A_\mu}{\partial x^\nu}. \tag{8.16}$$

We can not only earn money by trading money itself, but also by trading goods like, for example, copper. Depending on the local prices it can be lucrative to buy copper in one country, bring it to another country, sell it there, and then go back to the original country to compare the final amount of money with the amount of money we started with.

The gain factor for such a process is given by

$$g = \frac{p(\vec{n} + \vec{e}_i)}{p(\vec{n}) R_{\vec{n},i}}. \tag{8.17}$$

To understand this definition, imagine that we start with $10DM$ and the price for one kilogram of copper in Germany is $p(\vec{n}) = 10DM$. This means that we can buy exactly 1 kilogram of copper. Then we can go to the neighboring country and sell our copper for, say, $30F$ since $p(\vec{n} + \vec{e}_1) = 30F$. Afterward, we can go back to Germany and exchange our $30F$ for $15DM$ since, say, $R_{\vec{n},i} = 0.5$. Therefore, we have made in total $5DM$.

Again, a gain factor larger than one means that we earn money and a gain factor smaller than one that we lose money.[24]

[24] If you are unsure which quantity goes in the numerator and which in the denominator, ask yourself: Would it increase our profit if the given quantity is larger? If the answer is yes we write in the numerator, if not in the denominator. For example, a higher price of copper in France certainly increases our profit. Hence $p(\vec{n} + \vec{e}_i)$ is written in the numerator. Similarly, a higher price of copper in Germany would lower our profit and therefore, we write it in the denominator.

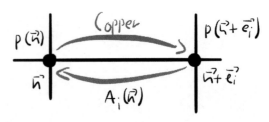

Again, we introduce the corresponding logarithm

$$g \equiv e^{J_i(\vec{n})} \tag{8.18}$$

and Eq. 8.17 then reads in terms of the corresponding logarithms

$$g = \frac{p(\vec{n}+\vec{e}_i)}{p(\vec{n})R_{\vec{n},i}}$$

$$e^{J_i(\vec{n})} = \frac{e^{\varphi(\vec{n}+\vec{e}_i)}}{e^{\varphi(\vec{n})}e^{A_i(\vec{n})}}$$

$$J_i(\vec{n}) = \varphi(\vec{n}+\vec{e}_i) - \varphi(\vec{n}) - A_i(\vec{n}). \tag{8.19}$$

The amount of money we earn depends on the amount of copper we carry around. Thus, in general, we have

$$J_i(\vec{n}) = q\Big(\varphi(\vec{n}+\vec{e}_i) - \varphi(\vec{n}) - A_i(\vec{n})\Big), \tag{8.20}$$

where q is the amount of copper involved in the trade. Completely analogous to what we did above, we can generalize this formula for situations that involve time by replacing $i \in \{1,2,\ldots,d\}$ with $\mu \in \{0,1,2,\ldots,d\}$).

$$J_\mu(\vec{n}) = q\Big(\varphi(\vec{n}+\vec{e}_\mu) - \varphi(\vec{n}) - A_\mu(\vec{n})\Big). \tag{8.21}$$

In addition to the interpretation as a gain factor, there is another way we can look at the four quantities $J_\mu(\vec{n})$.

As mentioned above, the amount of money we can earn in a copper trade J_μ is proportional to the amount of copper involved. Hence, we can use J_μ as a measure, for example, of the amount of copper that flows between countries. In the trade described by $J_i(\vec{n})$ copper is transported from the country at $\vec{n}$ to the neighboring country at $\vec{n}+\vec{e}_i$. Hence, $J_i(\vec{n})$ is a measure of the amount of copper that flows between the two countries.

The trade related to the gain factor $J_0(\vec{n})$ does not involve the exchange of copper between neighboring countries. Instead, $J_0(\vec{n})$ tells us how much money we can earn by buying copper and selling it at a later point in time in the *same* country. Again, $J_0(\vec{n})$ is directly proportional to the amount of copper involved. Hence, $J_0(\vec{n})$ is a measure of the amount of copper in the country at $\vec{n}$.

This is an important idea since so far we had nothing in our description that contained any information about the amount of copper at a certain location or about how copper flows through the system. Copper is represented in our description in somewhat abstract terms by its price. However, the amount of copper in a given country is not proportional to the local price of copper. The local price depends on the local money coordinate system and therefore cannot represent directly a quantity like the amount of copper. In other words, the amount of copper in a given country must be represented by a gauge independent quantity like $J_0(\vec{n})$. The same is true for the flow of copper $J_i(\vec{n})$.[25]

[25] In physics $J_0 = c\rho$ (see Section 2.2) describes the charge density while J_i describes the current density (see Section 2.3).

Now that we've introduced a mathematical notation to describe our toy model, we can also write down the specific rules according to which the system behaves.

8.3.2 Gauge rules

First of all, of course, it's possible to consider various kinds of dynamics within our financial toy model which correspond to different kinds of laws. However, in the following we discuss a very particular set of rules which correspond to what is known as Maxwell's equations in physics.

To derive these, we start with the crucial assumption that copper is conserved (i.e. no copper is destroyed or produced).[26] This means that whenever the amount of copper decreases in a given country it must have gone somewhere. Equally, whenever the amount of copper increases in a country it must have come from somewhere.

[26] By invoking Noether's first theorem and a Lagrangian formulation this can be derived using the global symmetry discussed above.

In words, this means:[27]

change of amount of copper in country $\vec{n}$ = total net flow

which we can write in mathematical terms as

$$J_0(\vec{n} + \vec{e}_0) - J_0(\vec{n}) = -\left(\sum_{i=1}^{d} J_i(\vec{n}) - \sum_{i=1}^{d} J_i(\vec{n} - \vec{e}_i)\right)$$

[27] We only consider elementary copper trade loops. In these elementary loops copper always flows "from left to right". For example, $J_i(\vec{n})$ is proportional to the amount of copper that flows from $\vec{n}$ to $\vec{n} + \vec{e}_i$. A positive $J_i(\vec{n})$ therefore represents an outflow of copper. Analogously, the flow involved in the process which yields $J_i(\vec{n} - \vec{e}_i)$ also goes from left to right. However, a positive $J_i(\vec{n} - \vec{e}_i)$ means a net inflow. Therefore, we need a relative minus sign. If we have a positive quantity on the left-hand side, the amount of copper increases and this must correspond to a net inflow.

where we sum over all neighboring countries. We can also write this as

$$J_0(\vec{n} + \vec{e}_0) - J_0(\vec{n}) + \left(\sum_{i=1}^{d} J_i(\vec{n}) - \sum_{i=1}^{d} J_i(\vec{n} - \vec{e}_i) \right) = 0. \qquad (8.22)$$

and we can visualize this equation as follows:

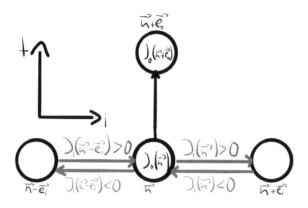

In the continuum limit, Eq. 1.4 becomes the usual **continuity equation**[28]

$$\sum_{\mu=0}^{3} \partial_\mu J_\mu = 0. \qquad (8.23)$$

[28] Here and in the following, we use the usual summation convention, i.e. there is an implicit sum over all indices that appear in pairs.

Now, what we really want is an equation that let's us understand how arbitrage opportunities show up and evolve as time passes. Moreover, we already know that the good quantities to describe our system are J_μ and $F_{\mu\nu}$ since these do not depend on local conventions. The quantities J_μ contain information about the positions and flow of copper, while $F_{\mu\nu}$ represent arbitrage opportunities.

However, the first naive guess to write $F_{\mu\nu}$ on one side and J_μ on the other side of an equation fails because J_μ has only one index but $F_{\mu\nu}$ has two. There is one additional piece of information that we can use, namely Eq. 8.23. If the right-hand side of an equation yields zero when we take the derivative $\partial_\mu J_\mu$, the left-hand side has, of course, to be zero too. But there is no reason why $\partial_\mu F_{\mu\nu}$ should be zero and this is another hint that our first naive guess is wrong.

The crucial observation is that

$$\sum_{\mu=0}^{3}\sum_{\nu=0}^{3} \partial_\nu \partial_\mu F_{\mu\nu} = 0, \tag{8.24}$$

since $F_{\mu\nu}$ is antisymmetric but partial derivatives commute $\partial_\mu \partial_\nu = \partial_\nu \partial_\mu$.[29] This suggests that we try

$$\sum_{\nu=0}^{3} \partial_\nu F_{\mu\nu} = \mu_0 J_\mu, \tag{8.25}$$

where we introduced the proportionality constant μ_0 which encodes how strongly the pattern of arbitrage opportunities react to the presence and flow of copper. This equation has exactly one free index (μ) on both sides and most importantly, both sides yield zero if we calculate the derivative

$$\sum_{\mu=0}^{3}\sum_{\nu=0}^{3} \partial_\mu \partial_\nu F_{\mu\nu} = \mu_0 \sum_{\mu=0}^{3} \partial_\mu J_\mu$$

$$0 = 0 \ \checkmark \qquad \qquad \text{Eq. 8.23 and Eq. 8.24}$$

[29] The sum over something antisymmetric times something symmetric is always zero. We discussed this explicitly in Chapter 7. This is analogous to how the integral over a symmetric function (e.g. $\cos(x)$) times an antisymmetric function (e.g. $\sin(x)$) over a symmetric interval yields exactly zero:

$$\int_{-a}^{a} \sin x \cos x = 0.$$

Eq. 8.25 is the famous **inhomogeneous Maxwell equation**.[30] For our discrete system it reads

$$\sum_{\nu=1}^{d} F_{\mu\nu}(\vec{n}) - \sum_{\nu=1}^{d} F_{\mu\nu}(\vec{n} - \vec{e}_\nu) = \mu_0 J_\mu(\vec{n}). \tag{8.26}$$

Moreover, Eq. 1.4 with $\mu = 0$ reads

$$\sum_{\nu=1}^{d} F_{0\nu}(\vec{n}) - \sum_{\nu=1}^{d} F_{0\nu}(\vec{n} - \vec{e}_\nu) = \mu_0 J_0(\vec{n})$$

$$\sum_{i=1}^{d} F_{0i}(\vec{n}) - \sum_{\nu=1}^{d} F_{0i}(\vec{n} - \vec{e}_i) = \mu_0 J_0(\vec{n}), \tag{8.27}$$

[30] Take note that this derivation of the inhomogeneous Maxwell equations is analogous to how the Einstein equation is often derived in textbooks. For an alternative derivation using a similar toy model, see [Maldacena, 2016].

where we used that $F_{00}(\vec{n}) = 0$, since $F_{\mu\nu}$ is antisymmetric. We have on the right-hand side $\mu_0 J_0(\vec{n})$, which is proportional to the amount of copper located at $\vec{n}$. Thus, for $\mu = 0$ Eq. 8.25

gives us information about the pattern of exchange rates around a country in which copper is present. In the continuum limit, Eq. 8.27 becomes

$$\sum_{i=0}^{3} \partial_i F_{0i} = \mu_0 J_0, \qquad (8.28)$$

which is known as **Gauss's law**.

Similarly, for $\mu = i \in \{1, 2, 3\}$, we get equations that give us information about the pattern of exchange rates which are present whenever copper flows

$$\sum_{\nu=1}^{d} F_{i\nu}(\vec{n}) - \sum_{\nu=1}^{d} F_{i\nu}(\vec{n} - \vec{e}_\nu) = \mu_0 J_i(\vec{n}). \qquad (8.29)$$

In the continuum limit, this equation becomes

$$\partial_0 F_{0i} + \sum_{j=0}^{3} \partial_j F_{ij} = \mu_0 J_i, \qquad (8.30)$$

which is known as the **Ampere-Maxwell law**.

There are two important lessons in this section. Firstly, we saw why redundancies can be useful. By thinking about redundancies we learn which quantities are independent of local conventions. Secondly, we derived the correct equation describing the interplay between goods like copper and arbitrage opportunities by using the ideas that we should describe the system with gauge invariant quantities and that copper is conserved.[31]

Now, we move on and discuss the simplest instances of gauge symmetries in physics.

8.4 Gauge symmetry in physics

8.4.1 Gauge symmetry in Quantum Mechanics

In our financial toy model, we used local prices $p(\vec{n})$ to describe copper. Analogously, in quantum mechanics we use the wave

[31] Take note that the homogeneous Maxwell equation

$$0 = \partial_\alpha \left(\frac{1}{2} \epsilon_{\alpha\beta\gamma\delta} F_{\delta\gamma} \right)$$
$$= \partial_\lambda F_{\mu\nu} + \partial_\mu F_{\nu\lambda} + \partial_\nu F_{\lambda\mu} \qquad (8.31)$$

follows directly from the definition of $F_{\mu\nu}$ (Eq. 8.14). This is discussed in Appendix A.16.1.

[32] This is a global transformation, analogous to, for example, a rotation of the whole subsystem. Under such a global rotation, all vectors have to be rotated accordingly. Analogously, all wave functions must be transformed, i.e. $\psi_i \to e^{i\epsilon}\psi_i$ too.

function $\Psi(x)$ to describe particles like, for example, an electron. A wave function is a complex function which can be written in polar form

$$\Psi(x) = R(x)e^{i\varphi(x)}. \tag{8.32}$$

Observables are related to products of the form $\psi_i^\star(x)\hat{O}\Psi(x)$, where $\hat{O}$ denotes an operator. Therefore, we can multiply the wave function by a global phase factor without changing anything[32]

$$\Psi(x) \to e^{i\epsilon}\Psi(x) \tag{8.33}$$

since

$$\psi_i^\star(x)\hat{O}\Psi(x) \to \psi_i^\star(x)e^{-i\epsilon}\hat{O}e^{i\epsilon}\Psi(x) = \psi_i^\star(x)\hat{O}\Psi(x) \tag{8.34}$$

It is important to note that this is an observable symmetry and is completely analogous to, for example, the rotational symmetry of Galileo's ship. To understand this, imagine that a quantum physicist sits inside the ship which represents our subsystem. This quantum physicist performs an experiment with electrons which are injected through a small slit. We can perform a global transformation as given in Eq. 8.33 by using a phase shifter. So we phase shift our electron before we insert it into the ship. The crucial point is that it's impossible for the quantum physicist inside the box to find out whether we performed such a phase shift or not. Therefore, a global phase shift is indeed a symmetry.

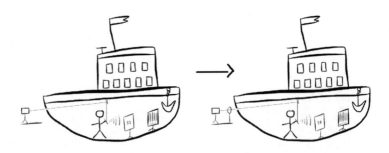

Now, what about local phase shifts in quantum mechanics? A local phase shift is a transformation of the form

$$\Psi(x) \to e^{i\epsilon(x)}\Psi(x), \tag{8.35}$$

where the transformation parameter $\epsilon(x)$ is now a function of the location x, i.e. no longer globally the same. Again it's important to keep in mind that this kind of transformation can be understood in an active and in a passive sense.

We can notice immediately that our description is not invariant since, for example, for the momentum operator $\hat{p} = -i\partial_x$ contains a derivative:[33]

[33] For simplicity, we restrict ourselves to one spatial dimension and work with $\hbar = 1$.

$$\psi_i^\star(x)\hat{p}\Psi(x) \to \psi_i^\star(x)e^{-i\epsilon(x)}\hat{p}e^{i\epsilon(x)}\Psi(x)$$

$$= -i\psi_i^\star(x)e^{-i\epsilon(x)}\partial_x e^{i\epsilon(x)}\Psi(x)$$

$$= -i\psi_i^\star(x)\partial_x\Psi(x) + \psi_i^\star(x)\left(\partial_x\epsilon(x)\right)\Psi(x)$$

$$\neq \psi_i^\star(x)\hat{p}\Psi(x). \tag{8.36}$$

However, if we interpret the local transformation in the passive sense, it shouldn't make any difference.[34] Analogous to what we did in our financial toy model, we can achieve this by introducing a bookkeeper A_μ which keeps track of such local changes of the phase. In particular, we replace the momentum operator with the so-called covariant momentum operator

[34] Reminder: passive transformation = coordinate transformation.

$$\hat{P} = -i\partial_x - A_x. \tag{8.37}$$

This bookkeeper A_x becomes under a local phase shift (Eq. 8.35)[35]

[35] This is exactly the continuum limit of Eq. 8.10.

$$A_x \to A_x + \partial_x\epsilon(x). \tag{8.38}$$

After the introduction of this bookkeeper our description is indeed invariant under local phase shifts.

$$\psi_i^\star(x)\hat{P}\Psi(x) \to \psi_i^\star(x)e^{-i\epsilon(x)}\hat{P}e^{i\epsilon(x)}\Psi(x)$$

$$= \psi_i^\star(x)e^{-i\epsilon(x)}\left(-i\partial_x - A_x - \partial_x\epsilon(x)\right)e^{i\epsilon(x)}\Psi(x)$$

$$= -i\psi_i^\star(x)\partial_x\Psi(x) + \psi_i^\star(x)\left(\partial_x\epsilon(x)\right)\Psi(x)$$

$$- \psi_i^\star(x)A_x\Psi(x) - \psi_i^\star(x)\left(\partial_x\epsilon(x)\right)\Psi(x)$$

$$= \psi_i^\star(x)\hat{P}\Psi(x). \tag{8.39}$$

But are local phase shifts also a real symmetry of quantum mechanics?

To understand this, we again ask our quantum physicist in the ship to perform an experiment with electrons. However, this time we do not perform the phase shifts globally but locally. This means that we put the phase shifter *inside* the ship.

Now, the quantum physicist can find out that a phase shift did happen, even if he can't see the phase shifter directly. All he has to do is perform a double slit experiment. The result of the double slit experiment is dramatically altered by a local phase shift[36].

[36] Y. Aharonov and D. Bohm. Significance of electromagnetic potentials in the quantum theory. *Phys. Rev.*, 115:485–491, 1959. DOI: 10.1103/PhysRev.115.485. [,95(1959)]; and Gerard 't Hooft. Gauge theories of the forces between elementary particles. *Sci. Am.*, 242N6:90–116, 1980. [,78(1980)]

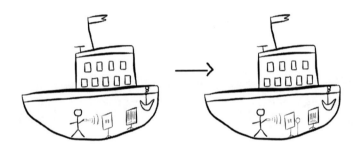

Therefore, we can conclude that local phase shifts are not symmetries of quantum mechanics.

Now, it's again possible that our bookkeepers A_μ become real dynamical actors and are completely analogous to what we discussed for our financial toy model. The theory which describes the dynamics of the bookkeepers is known as electrodynamics.

8.4.2 Gauge Symmetry in Electrodynamics

The equations of electrodynamics (Maxwell's equations) also posses a global symmetry

$$A_\mu(x) \to A_\mu(x) + a_\mu, \tag{8.40}$$

where a_μ are arbitrary real numbers. This comes about since we can only measure potential differences.[37]

[37] We discussed this in the previous chapter.

Like in the two previous examples, the invariance under this global transformation is a real observable symmetry. To understand this, we again imagine a physicist in a ship which this time, however, is isolated from the ground.

Charging the ship leads to a global increase of the electric potential $A_0(x) \to A_0(x) + \phi_0$, while $A_i(x)$ remains unchanged. However, this change has no measurable effect inside the subsystem since $B_i = \epsilon_{ijk}\partial_j A_k$ and $E_i = -\partial_i A_0 - \partial_t A_i$ are unchanged.

So, as long as we raise the electric potential A_0 globally inside the ship, there is no possibility for the physicist to detect that we changed the potential.[38]

[38] If this is unclear, recall that, for example, a bird can walk on an uninsulated power line without any injuries. This is possible because the bird walks solely in the subsystem "power line" and the electric potential is only high relative to the potential at the ground. Alternatively, imagine Faraday sitting in a metal cage which is insulated from the ground. It's possible to electrify the cage without any notable effect inside the cage. In other words, we can raise the electric potential globally inside such a Faraday cage without inducing any measurable effect [Aharonov and Bohm, 1959]. We can understand this by recalling that in Section 4.1.2, we discovered that the electric field inside a charged sphere is zero.

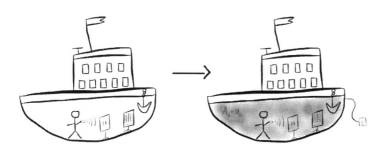

Similarly to what we discussed above, we can also argue that a local shift of the electromagnetic potential is not a symmetry of the system. To understand this, imagine that we only change the potential of a single object inside the ship. Clearly the physicist would have no problem finding this out.

The most important point is that in electrodynamics, our bookkeepers A_μ are dynamical physical actors and they can induce real physical changes. A famous example of this phenomenon is the so-called Aharonov-Bohm effect, where a non-zero A_μ induces a phase shift in the wave function.[39]

[39] Y. Aharonov and D. Bohm. Significance of electromagnetic potentials in the quantum theory. Phys. Rev., 115:485–491, 1959. DOI: 10.1103/PhysRev.115.485. [,95(1959)]

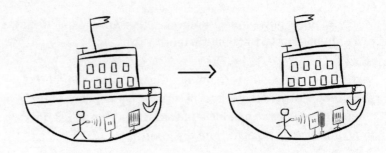

Moreover, completely analogous to what we did in Eq. 8.14 we can define the quantity[40]

$$F_{\mu\nu}(x_\mu) \equiv \frac{\partial A_\nu}{\partial x^\mu} - \frac{\partial A_\mu}{\partial x^\nu}, \tag{8.41}$$

which encodes in gauge invariant terms information about the presence of electromagnetic fields. In the context of electrodynamics, the quantity in Eq. 8.41 is known as the **field strength tensor**.

[40] In our financial toy model, this quantity encodes how much money we can make by trading money in a loop (i.e. about an arbitrage opportunity).

With this in mind, we can now put the puzzle pieces together and disentangle real symmetries from mere redundancies in quantum mechanics and electrodynamics.

8.4.3 Putting the puzzle pieces together

First, we noted that there is a global symmetry in quantum mechanics (Eq. 8.33)

$$\Psi(x) \to e^{i\epsilon}\Psi(x), \tag{8.42}$$

but not a local one (Eq. 8.35)

$$\Psi(x) \to e^{i\epsilon(x)}\Psi(x). \tag{8.43}$$

However, we then learned that we can rewrite our equations such that they are invariant under local transformations by introducing bookkeepers A_μ.

We then argued that the invariance under global transformations represents a real symmetry, while the invariance under local transformations is only a redundancy. Formulated differently, we have invariance under active global transformations and also passive local transformations if we formulate the theory appropriately. However, quantum mechanics is not invariant under active local phase shifts.

Analogous to what we did in the financial toy model, we then argued that our bookkeepers A_μ can also appear as real dynamical parts of the system and not only as purely mathematical bookkeepers. The theory which describes the dynamics of the bookkeepers is electrodynamics. The bookkeepers A_μ then become what we usually call the electromagnetic potential.

The crucial point is then that as soon as we have a system where the bookkeepers are no longer purely mathematical objects but physical parts of it, they can induce measurable changes. An important example is the Aharonov-Bohm effect, where a nonzero potential induces a phase shift in the wave function[41].

[41] Y. Aharonov and D. Bohm. Significance of electromagnetic potentials in the quantum theory. Phys. Rev., 115:485–491, 1959. DOI: 10.1103/PhysRev.115.485. [,95(1959)]

What this implies immediately is that, in principle, it's *possible* to cancel any local phase shift using an electromagnetic potential A_μ. However, it's important to take note that we still do not have a local symmetry when an electromagnetic potential is present, for example, in the ship in which our quantum physicist detects local phase shifts.

It is instructive to reformulate this point using the notions of "active transformation" and "passive transformation" which were introduced in Section 8.2.1.

A passive transformation is simply a change of the coordinate system and therefore cannot lead to any physical change. All we achieve through a passive transformation is a different description of the same physical situation. Therefore, when we perform a passive transformation, we must be careful to keep our description consistent. In particular, this means that whenever we perform a local passive transformation, we have to accompany

$$\Psi \to e^{i\epsilon(x)}\Psi \qquad (8.44)$$

always with
$$A_\mu \to A_\mu + \partial_\mu \epsilon(x). \tag{8.45}$$

When we perform a passive transformation these two transformations always go hand in hand.

In contrast, an active transformation means that a real physical change happens and therefore the physical situation doesn't need to remain unchanged. Especially, when we induce an active local phase shift
$$\Psi \to e^{i\epsilon(x)}\Psi, \tag{8.46}$$

the corresponding transformation of A_μ does not happen automatically. Otherwise we wouldn't be able to detect local phase shifts in experiments. It *is* possible to cancel the phase shift in Ψ by using A_μ. But this only happens when we actively prepare the system in a particular way, for example, by using an Aharonov-Bohm type setup. In other words, the active shifts in Ψ and A_μ are two separate transformations which can happen independently.

To summarize: quantum mechanics and electrodynamics are invariant under active global gauge transformations. Hence, the global gauge symmetry is a real symmetry. However, only our description is invariant under passive local transformations. Therefore, local gauge symmetry is a redundancy.

Now, after these discussions of gauge symmetries in intuitive and concrete physical terms, it's time to move on and discuss how we can define them mathematically.

8.5 Gauge symmetries mathematically

First of all, symmetries are described mathematically using group theory. A group consists of all transformations which leave the given system invariant and an operation which allows us to connect transformations.[42]

[42] For further details, see, for example, [Schwichtenberg, 2018a]

In our financial toy model, the global symmetry group is the dilation group which consists of all possible dilations

$$f = e^\epsilon, \quad \text{with} \quad \epsilon \in \mathbb{R}. \tag{8.47}$$

of the given currency. The mathematical name for this group is $GL^+(1, R)$, the one-dimensional real general linear group with positive determinant.

In electrodynamics, the global symmetry group is $U(1)$ and consists of all possible phase shifts

$$f = e^{i\epsilon}, \quad \text{with} \quad \epsilon \in \mathbb{R}. \tag{8.48}$$

The difference between $GL^+(1, R)$ and $U(1)$ is the factor of i in the exponent of the transformation operators.[43]

[43] As a result, $U(1)$ is compact while $GL^+(1, R)$ is not.

An important idea was then to introduce bookkeepers A_μ in order to make the theory locally redundant. These bookkeepers allow us to use arbitrary local coordinate systems. In mathematical terms, we then have a local gauge symmetry, which is really just a redundancy in our description. So after the introduction of the bookkeepers, we can perform local dilations[44]

$$f(x) = e^{\epsilon(x)}, \quad \text{with} \quad \epsilon(x) \in C^\infty \tag{8.49}$$

[44] To unclutter the notation, we restrict ourselves to one spatial dimension.

since our bookkeepers adjust accordingly (Eq. 8.15). Analogously, in electrodynamics we can then perform local phase shifts (Eq. 8.35)

$$f(x) = e^{i\epsilon(x)}, \quad \text{with} \quad \epsilon(x) \in C^\infty. \tag{8.50}$$

The crucial point is that our transformation parameters $\epsilon(x)$ are functions of the location x. In other words, we can now shift the prices, or analogously the phase, at each point in space x by a different amount. Only global shifts were permitted without the bookkeepers which meant that the prices, or analogously the phases, were shifted by exactly the same amount everywhere.

After the introduction of the bookkeepers, we have the freedom to perform independent $GL^+(1, R)$ transformations at each point in space. In our financial toy model, this means that we have a copy of the **gauge group** $GL^+(1, R)$ at each point in space. Taken together these copies yield the **group of gauge transformations**.

Analogously, in electrodynamics our symmetry group is $U(1)$ and we also have a copy of $U(1)$ above each point in space.[45] Since each Lie group can be understood as a manifold, the geometrical picture involving fiber bundles emerges. For example, $U(1)$ transformations (Eq. 8.33) are unit complex numbers, which all lie on the unit circle in the complex plane. Hence, geometrically we can imagine $U(1)$ as a circle and therefore that there is a little circle attached to each spacetime point.

[45] It's crucial to keep the notions of a "gauge group" and a "group of gauge transformations" separate. While the former is usually simple and finite-dimensional, the latter is a lot more complicated and infinite-dimensional. This comes about since the group of gauge transformations is a group of smooth functions on spacetime that take values in the gauge group.

Since pictures with lots of circles quickly become confusing, it is conventional to "cut" the circle and turn it into a line. We then have to remember that the end points P have to be identified.

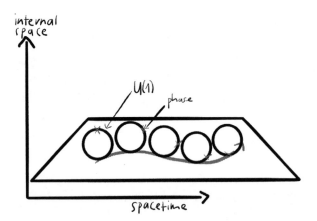

The final geometrical picture is known as a **fiber bundle**:

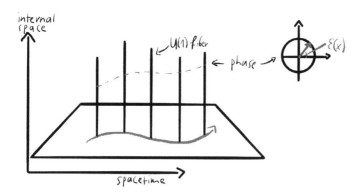

A final thing that we need to talk about are our bookkeepers, which we introduced to make the theory locally invariant. In mathematical terms the bookkeepers A_μ are called **connections**. A connection is a tool that allows us to compare prices or phases at different locations since it keeps track of how the local coordinate systems are defined and encodes information about the structure of the space we are moving on.

As already mentioned above, there are two situations where it's necessary to introduce connections. On the one hand, we need connections to allow for arbitrary local coordinate systems. Here, the connection keeps track of these local coordinate systems and lets us compare prices or phases defined according to different local conventions. On the other hand, connections are essential whenever the space we are interested in is curved.

The prototypical example of a curved space is a sphere. To compare vectors at two different points on a sphere (e.g. to calculate a derivative), we need a procedure to move one vector to the location of the second one consistently.

The needed procedure is known as **parallel transport**. To understand it, imagine that you are walking on the sphere while holding a stick in your hand. While you walk, you always do your best to keep the stick straight. If you do this, you are parallel transporting the stick.

Mathematically, the infinitesimal parallel transport of a vector $V_\alpha(x)$ is defined as

$$V_\alpha(x+dx) = V_\alpha(x) - \Gamma^\alpha_{\beta\gamma}(x) V^\beta(x) dx^\gamma, \qquad (8.51)$$

where Γ^i_{jk} denotes the corresponding connection.

An important observation is that if the space you are moving in is curved, it's possible that the stick does not end up in its starting position if you move along a closed curve.

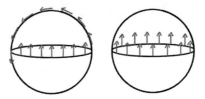

Hence, the difference between the original vector and the vector which was parallel transported along an infinitesimal closed curve encodes information about the local curvature. Therefore, we imagine that our vector moves from A to B via two different paths.

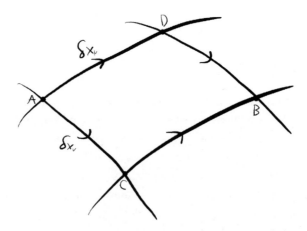

Taken together these two paths yield a closed curve and we can calculate

$$V_\alpha(A \to C \to B) - V_\alpha(A \to D \to B) = R_{\alpha\beta\ \mu}^{\ \ \nu} V^\beta dx^\mu dx^\nu + \ldots, \tag{8.52}$$

where $R_{\alpha\beta\ \mu}^{\ \ \nu}$ denotes the corresponding (Riemann) curvature tensor

$$R_{\alpha\beta\ \mu}^{\ \ \nu} = \partial_\alpha \Gamma_{\beta\nu}^{\mu} - \partial_\beta \Gamma_{\alpha\nu}^{\mu} + \Gamma_{\alpha\kappa}^{\mu}\Gamma_{\beta\nu}^{\kappa} - \Gamma_{\beta\kappa}^{\mu}\Gamma_{\alpha\nu}^{\kappa}. \tag{8.53}$$

Now, it is also possible that we have curvature in an internal space. For example, for the $U(1)$ symmetry described above, we can imagine that the various $U(1)$ copies are glued together non-trivially.

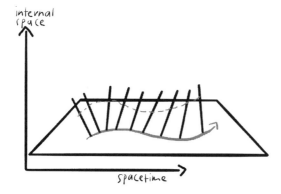

In this case, we again need a connection that allows us to consistently move our wave function from one point to another

$$\Psi(x_\mu + \Delta x_\mu) = \Psi(x_\mu) - A_\mu(x_\mu)\Psi(x_\mu)\Delta x_\mu, \tag{8.54}$$

where $A_\mu(x_\mu)$ denotes the connection. Moreover, completely analogously, we can imagine that we don't end up with the same wave function when we move along a closed curve. If this is the case, we know that our internal space is curved and hence, we define

$$F_{\mu\nu}(x_\mu) \equiv \frac{\partial A_\nu}{\partial x^\mu} - \frac{\partial A_\mu}{\partial x^\nu} \tag{8.55}$$

as a measure of the curvature. Take note that this is exactly how we defined the quantity which encodes information about arbitrage opportunities in Eq. 8.14 and the quantity which tells us that there is a non-zero electromagnetic field in Eq. 8.41.

An important point is that connections can be non-zero, even though the curvature is zero. In this case, our connections are necessary only because of our choice of the local coordinate systems and not as a result of the physical situation itself. If the curvature is zero, it is possible to find a choice of local coordinate systems such that no connection is necessary. However, whenever the curvature is non-zero, connections are essential and can't be removed by a clever choice of local coordinate systems.

This will be discussed further in the next section.

8.5.1 Gauge connections in Quantum Mechanics and the toy model

We can modify our equations in quantum mechanics to make them invariant under local $U(1)$ transformations. To achieve this, we need to introduce the connection A_μ. However, this connection is again a purely mathematical bookkeeper, as long as the curvature tensor vanishes[46]

$$F_{\mu\nu}(x_\mu) = 0, \tag{8.56}$$

where (Eq. 8.16)

$$F_{\mu\nu}(x_\mu) \equiv \frac{\partial A_\nu}{\partial x^\mu} - \frac{\partial A_\mu}{\partial x^\nu}. \tag{8.57}$$

But if $F_{\mu\nu} \neq 0$, we can't get rid of A_μ everywhere at the same time by a change of coordinate systems, since this would imply $F_{\mu\nu} = 0$, which corresponds to a different physical situation. Moreover, $F_{\mu\nu}$ only becomes the dynamical actor that we call electromagnetic field if there are nontrivial equations of motion for the connection (the Maxwell equations).

Finally, we can also come back to our financial toy model. Here, we can introduce arbitrary local currencies and this makes it

---

[46] Here curvature is not referring to a property of spacetime but of our internal space.

necessary to introduce bookkeepers $A_\mu(\vec{n})$ which are able to handle the exchange of currencies. However, these bookkeepers are purely mathematical parts of our description as long as there is no arbitrage opportunity

$$F_{\mu\nu}(\vec{n}) = 0 \qquad (8.58)$$

where (Eq. 8.14)

$$F_{\mu\nu}(\vec{n}) = A_\nu(\vec{n} + \vec{e}_\mu) - A_\nu(\vec{n}) - [A_\mu(\vec{n} + \vec{e}_\nu) - A_\mu(\vec{n})], \qquad (8.59)$$

The bookkeepers are only indispensable when there are real arbitrage opportunities $F_{\mu\nu}(\vec{n}) \neq 0$. Moreover, as soon as there are equations of motion for them, they become dynamical actors.

So in summary, while connections also appear when we write down the equations of a given model in a more general way, they have no measurable effect since we are still describing the same model. The physical situation is only then a different one when the corresponding curvature is non-zero. In this case, our connections are no longer optional but essential parts of the model and have a measurable effect on the dynamics. Moreover, they become dynamical actors only when they change dynamically, i.e. they follow their own equations of motion. Theories with a dynamical connection are what we call **gauge theories**. Since in electrodynamics we have Maxwell's equations for the connection A_μ and, in general, a non-zero curvature $F_{\mu\nu}$, we can understand electrodynamics as a gauge theory.

9

Further Reading Recommendations

As mentioned already in the preface, the content of this book is far from comprehensive. There are hundreds of different aspects of electrodynamics that I haven't even said a word about. Electrodynamics is almost two centuries old and thousands of people have worked on it. Unsurprisingly, no single book can capture it all.

However, there are lots of excellent books that cover various aspects extremely well. Since for every good book there are at least 20 bad ones which are not worth your time, I recommend some of my favorites below. So, start by picking the ones that interest you most, and dig in[1].

My favorite student-friendly electrodynamics textbooks are

▷ **Vol. 2 of the Feynman Lectures**[2]

▷ **A student's guide to Maxwell's equations** by Daniel A. Fleisch[3]

[1] If you need further or more specialized reading recommendations, you should visit:
www.physicstravelguide.com
This is an expository physics wiki where anyone can help to collect the best resources on any physics topic + publish student-friendly explanations.

[2] Richard Feynman. *The Feynman lectures on physics*. Addison-Wesley, San Francisco, Calif. Harlow, 2011. ISBN 9780805390650

[3] Daniel Fleisch. *A student's guide to Maxwell's equations*. Cambridge University Press, Cambridge, UK New York, 2008. ISBN 978-0521701471

▷ **Introduction to Electrodynamics** by David J. Griffiths[4]

▷ **Electricity and Magnetism** by Edward M. Purcell[5]

If you want to dive deeper, you might try

▷ **Modern Electrodynamics** by Andrew Zangwill[6]

▷ **Classical Electrodynamics** by John David Jackson[7]

However, be warned that these books aren't as student-friendly as the books mentioned above.

To learn more about vector and tensor calculus, I strongly recommend

▷ **Div, Grad, Curl, and All that** by H. M Schey[8]

▷ **A Student's Guide to Vectors and Tensors** by Daniel A Fleisch[9]

▷ **The vector calculus series** by Kalid Azad[10]

Good books to learn more about special relativity are

▷ **Special Relativity** by Anthony French[11]

▷ **Special Relativity for Beginners** by Jürgen Freund[12]

▷ **Spacetime Physics** by Edwin F. Taylor and John A. Wheeler[13]

If you want to learn more about gauge theory and group theory, you might enjoy my first book

▷ **Physics from Symmetry**[14]

---

[4] David Griffiths. *Introduction to electrodynamics*. Pearson Education Limited, Harlow, 2014. ISBN 9781292021423

[5] Edward Purcell. *Electricity and magnetism*. Cambridge University Press, Cambridge, 2013. ISBN 9781107014022

[6] Andrew Zangwill. *Modern electrodynamics*. Cambridge University Press, Cambridge, 2013. ISBN 9780521896979

[7] John Jackson. *Classical electrodynamics*. Wiley, New York, 1999. ISBN 9780471309321

[8] H. M. Schey. *Div, grad, curl, and all that : an informal text on vector calculus*. W.W. Norton & Company, New York, 2005. ISBN 9780393925166

[9] Daniel Fleisch. *A student's guide to vectors and tensors*. Cambridge University Press, Cambridge New York, 2012. ISBN 9781139031035

[10] https://betterexplained.com/articles/category/math/vector-calculus/

[11] A. P. French. *Special relativity*. Norton, New York, 1968. ISBN 9780393097931

[12] Juergen Freund. *Special relativity for beginners : a textbook for undergraduates*. World Scientific, Singapore, 2008. ISBN 9789812771599

[13] Edwin Taylor. *Spacetime physics : introduction to special relativity*. W.H. Freeman, New York, 1992. ISBN 9780716723271

[14] Jakob Schwichtenberg. *Physics from Symmetry*. Springer, Cham, Switzerland, 2018a. ISBN 978-3319666303

To learn more about fiber bundles and differential geometry, good starting points are

▷ **Fiber Bundles and Quantum Theory** by Herbert J. Bernstein and Anthony V. Phillips[15]

▷ **A pictorial introduction to differential geometry, leading to Maxwell's equations as three pictures** by Jonathan Gratus[16]

If you're looking for an introductory book on quantum mechanics you might enjoy my book

▷ **No-Nonsense Quantum Mechanics**[17]

[15] H. J. Bernstein and A. V. Phillips. Fiber Bundles and Quantum Theory. *Sci. Am.*, 245:94–109, 1981. DOI: 10.1038/scientificamerican0781-122

[16] J. Gratus. A pictorial introduction to differential geometry, leading to Maxwell's equations as three pictures. *ArXiv e-prints*, September 2017

[17] Jakob Schwichtenberg. *No-Nonsense Quantum Mechanics*. No-Nonsense Books, Karlsruhe, Germany, 2018b. ISBN 978-1719838719

One Last Thing

It's impossible to overstate how important reviews are for an author. Most book sales, at least for books without a marketing budget, come from people who find books through the recommendations on Amazon. Your review helps Amazon figure out what types of people would like my book and makes sure it's shown in the recommended products.

I'd never ask anyone to rate my book higher than they think it deserves, but if you like my book, please take the time to write a short review and rate it on Amazon. This is the biggest thing you can do to support me as a writer.

Each review has an impact on how many people will read my book and, of course, I'm always happy to learn about what people think about my writing.

PS: If you write a review, I would appreciate a short email with a link to it or a screenshot to Jakobschwich@gmail.com. This helps me to take note of new reviews. And, of course, feel free to add any comments or feedback that you don't want to share publicly.

Part IV
Appendices

A

Vector Calculus

A whole new set of mathematical tools is necessary to describe electrodynamics. Most of them are part of what is usually called vector calculus and this is what this appendix is about. However, we will not talk about vector calculus in general but instead restrict ourselves to those tools which are necessary to understand electrodynamics.[1]

[1] The only exception is the outer product which is only mentioned for completeness.

There are two ways of approaching this appendix. One possibility is to read it before you start reading the main chapters about electrodynamics. Alternatively, it is also possible to simply start reading the chapters about electrodynamics and then read about the concepts whenever we really need them.

As usual, we'll start with a birds-eye overview and afterwards dive into the details.

Basic Quantities

To describe electrodynamics, we need more than simple numbers.[2] For example, charged objects get pushed by other charged objects in a particular direction. For this reason, we need **vectors** $\vec{v}$, which not only allow us to describe how strongly an object gets pushed but also in which direction. Moreover, to get a deep understanding of electrodynamics, we need to understand the interplay between the electric and the magnetic field. This interplay can be best understood by making use of a particular **tensor** $F_{\mu\nu}$, which allows us to describe more than one direction at once.[3]

In addition, since charged objects can influence each other without directly touching each other, we need to introduce so-called **fields**. A field is a mathematical object which assigns a particular quantity (scalar, vector, tensor) to each point in space. In particular, the electric field $\vec{E}(\vec{x})$ and magnetic field $\vec{B}(\vec{x})$ are mathematically **vector fields**. They describe at each location $\vec{x}$ a particular field strength and direction in which a charged object would get pushed. Moreover, the electromagnetic field tensor $F_{\mu\nu}(\vec{x})$ assigns a tensor to each location $\vec{x}$ and is therefore mathematically a tensor field.[4]

[2] In mathematical terms a simple number is a **scalar**. The defining feature of a scalar is that it doesn't change if we change our coordinate system while, for example, a vector gets rotated if we rotate our coordinate system.

[3] Scalars, vectors and tensors are discussed in Appendix A.1

[4] We talk about fields in Appendix A.4. Take note that in physics our fields usually also change in time, i.e. $F_{\mu\nu} = F_{\mu\nu}(t,\vec{x})$ and $\vec{E} = \vec{E}(t,\vec{x})$ etc.

Basic Operations

There are different ways how we can combine vectors and vector fields.[5] One possibility is to multiply two vectors $\vec{v}$ and $\vec{w}$ and, as a result, get a matrix

$$\vec{v} \otimes \vec{w} \equiv \vec{v}\vec{w}^T = \begin{pmatrix} v_1 \\ v_2 \\ v_3 \end{pmatrix} \begin{pmatrix} w_1 & w_2 & w_3 \end{pmatrix}$$

$$= \begin{pmatrix} v_1 w_1 & v_1 w_2 & v_1 w_3 \\ v_2 w_1 & v_2 w_2 & v_2 w_3 \\ v_3 w_1 & v_3 w_2 & v_3 w_3 \end{pmatrix}.$$

[5] The reason why we care about these products is, well, that we need them to describe nature.

This way of combining two vectors is known as the **outer product**.

Another closely related possibility to combine two vectors is

$$\vec{v} \cdot \vec{w} \equiv \vec{v}^T \vec{w} = \begin{pmatrix} v_1 & v_2 & v_3 \end{pmatrix} \begin{pmatrix} w_1 \\ w_2 \\ w_3 \end{pmatrix}$$

$$= v_1 w_1 + v_2 w_2 + v_3 w_3.$$

Here the result is a simple number and this way of combining two vectors is known as the **inner product** or the **scalar product** or the **dot product**.[6]

[6] The dot product is the topic of Appendix A.2.

A third product allows us to combine two vectors in such a way that we get a vector as a result:

$$\vec{v} \times \vec{w} = \begin{pmatrix} v_1 \\ v_2 \\ v_3 \end{pmatrix} \times \begin{pmatrix} w_1 \\ w_2 \\ w_3 \end{pmatrix}$$

$$= \begin{pmatrix} v_2 w_3 - v_3 w_2 \\ v_3 w_1 - v_1 w_3 \\ v_1 w_2 - v_2 w_1 \end{pmatrix}.$$

This way of combining two vectors is known as the **cross product**.[7]

[7] The cross product is discussed in Appendix A.3.

To summarize:

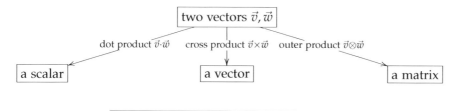

Integrals

There are different kinds of integrals for fields. The simplest type is a **line integral**[8]

[8] See Appendix A.5.

$$\int_C \phi(\vec{r})dl, \qquad (A.1)$$

where we integrate a scalar field $\phi(\vec{r})$ along some curve C.

We need to be a bit more careful when we integrate a vector field along some path P

$$\int_P \vec{V}(\vec{x}) \cdot d\vec{l} = \int_P \vec{V} \cdot \vec{t}(l)dl, \qquad (A.2)$$

since we need to take the direction of $\vec{V}(\vec{x})$ at each location $\vec{x}$ into account. This type of integral is known as a **path integral**.[9]

[9] The path integral is the topic of Appendix A.6.

In addition, we can consider higher-dimensional integrals like a **surface integral**[10]

[10] See Appendix A.8.

$$\int_S \phi(\vec{x})da, \qquad (A.3)$$

where a scalar field $\phi(\vec{r})$ is integrated over some surface S.

The analogous integral for vector fields is again a bit more complicated

$$\int_S \vec{V}(\vec{x}) \cdot d\vec{a} = \int_S \vec{V}(\vec{x}) \cdot \vec{n}da, \qquad (A.4)$$

since again, we need to take the direction of $\vec{V}(\vec{x})$ at each location $\vec{x}$ into account. A surface integral for a vector field is usually called a **flux integral**.[11]

[11] We discuss how this name comes about in detail in Appendix A.9.

Differential Operators

There are different ways of gathering information about how a given field changes through space. For an ordinary function $f(x)$ this kind of information is encoded in the derivative $\partial_x f(x)$. For a scalar field $\phi(\vec{x})$, we need to describe how it changes as we move in the three spatial directions (x, y, z):

$$\nabla \phi(\vec{x}) = \begin{pmatrix} \partial_x \\ \partial_y \\ \partial_z \end{pmatrix} \phi(\vec{x}) = \begin{pmatrix} \partial_x \phi(\vec{x}) \\ \partial_y \phi(\vec{x}) \\ \partial_z \phi(\vec{x}) \end{pmatrix}. \qquad (A.5)$$

This is known as the **gradient** of $\phi(\vec{x})$.[12]

Moreover, vector fields $\vec{F}(\vec{x})$ additionally assign a direction to each point in space. Information about how this direction changes as we move through space is encoded in the **divergence**[13]

$$\nabla \cdot \vec{F}(\vec{x}) = \begin{pmatrix} \partial_x \\ \partial_y \\ \partial_z \end{pmatrix} \cdot \begin{pmatrix} F_x(x) \\ F_y(x) \\ F_z(x) \end{pmatrix} = \partial_x F_x(\vec{x}) + \partial_y F_y(\vec{x}) + \partial_z F_z(\vec{x}).$$

(A.6)

and in the **curl**[14]

$$\nabla \times \vec{F}(\vec{x}) = \begin{pmatrix} \partial_x \\ \partial_y \\ \partial_z \end{pmatrix} \times \begin{pmatrix} F_x(\vec{x}) \\ F_y(\vec{x}) \\ F_z(\vec{x}) \end{pmatrix} = \begin{pmatrix} \partial_y F_z(\vec{x}) - \partial_z F_y(\vec{x}) \\ \partial_z F_x(\vec{x}) - \partial_x F_z(\vec{x}) \\ \partial_x F_y(\vec{x}) - \partial_y F_x(\vec{x}) \end{pmatrix}. \quad (A.7)$$

[12] We discuss the gradient in Appendix A.10.

[13] The divergence is the topic of Appendix A.11.

[14] The meaning of the curl of a vector field is discussed in detail in Appendix A.12.

The divergence tells us if the arrows around a given location point in the same direction or away from the location. The curl tells us whether or not the arrows circle around the location.

It is important to keep in mind that many quantities like divergence and curl were specifically introduced because we need them in electrodynamics. In other words, the main motivation for why these quantities are introduced is that we need them to describe nature.

Fundamental Theorems

There are several theorems which allow us to understand the interplay between the various differential operators (gradient, divergence, curl) and the various types of integrals (line integral, path integral, surface integral, flux integral).

The basic message in all these theorems is that the integral of a derivative over a region is equal to the sum over the values of the quantity at the boundary of the region.[15]

[15] Take note that the same is true for the fundamental theorem of calculus $\int_a^b \frac{df(x)}{dx} dx = f(b) - f(a)$. Here our region is the line from a to b. The theorem allows us to replace the integral over this region by a sum over the values of $f(x)$ at the boundary of the line, i.e. a and b.

For example, the **fundamental theorem for gradients**

$$\int_P \vec{\nabla}\phi(\vec{x}) \cdot d\vec{s} = \phi(\vec{r}_1) - \phi(\vec{r}_0), \quad (A.8)$$

tells us that the path integral over the gradient of the scalar field $\phi(\vec{x})$ is equal to the difference of the values of $\phi(\vec{x})$ at the boundary of the path P.[16]

[16] The fundamental theorem for gradients is the topic of Appendix A.13. The boundary of a path is the starting and the end point of the path ($\vec{r}_0$ and $\vec{r}_1$).

Analogously, the **fundamental theorem for divergences**[17]

$$\int_V \vec{\nabla} \cdot \vec{F}(\vec{x}) dV = \oint_S \vec{F}(\vec{x}) \cdot d\vec{S} \quad (A.9)$$

tells us that the volume integral over the divergence of a vector field $\vec{F}(\vec{x})$ is equal to the sum over the values of $\vec{F}(\vec{x})$ at the boundary of V.[18]

[17] The fundamental theorem for divergences is also known as Gauss's theorem, Green's theorem or simply the divergence theorem. We talk about it in detail in Appendix A.13.

[18] The boundary of a volume V is what we call its surface S. Moreover, recall that an integral can be understood as a finely-grained sum. This means that on the right-hand side, we really sum over all values of the vector field $\vec{F}(\vec{x})$ on the surface S.

Finally, the **fundamental theorem for curls**[19]

$$\int_S \vec{\nabla} \times \vec{F}(\vec{x}) \cdot d\vec{S} = \oint_P \vec{F}(\vec{x}) \cdot d\vec{l} \quad (A.10)$$

tells us that the surface integral over the curl of a vector field $\vec{F}(\vec{x})$ is equal to the sum over the values of $\vec{F}(\vec{x})$ at the boundary of S.[20]

[19] The fundamental theorem for curls is also known as Stokes' theorem. We discuss it in detail in Appendix A.15.

[20] The boundary of a surface S is a path P around it.

Now, let's talk about these concepts in detail.

A.1 Scalars, Vectors, Tensors

In somewhat naive terms, we can say that a scalar is simply a number, a vector an arrow and a tensor something more complicated like a matrix.

$$\text{scalar} \qquad \text{vector} \qquad \text{rank 2 tensor}$$

$$(x) \qquad \begin{pmatrix} x \\ y \\ z \end{pmatrix} \qquad \begin{pmatrix} M_{11} & M_{12} & M_{13} \\ M_{21} & M_{22} & M_{23} \\ M_{31} & M_{32} & M_{33} \end{pmatrix}$$

We use scalars to describe quantities that only have a magnitude (or size) and no associated direction, for example, like temperature. In contrast, a vector is a mathematical object that we use whenever we want to describe something which is characterized by its magnitude *and* direction. An example of a quantity that we need to describe using a vector is the velocity of an object. We need a vector here since we not only need to describe how fast the object is moving but also in which direction. Another important example of a vector quantity is a force. Here, we describe how strongly something gets pushed in a particular direction.

A tensor is an object we use to describe quantities which are characterized by their magnitudes and *multiple* directions. The number of directions necessary determines the rank of the tensor. For example, if two directions are needed, we need a rank 2 tensor.[21] One of the most important tensors is the electromagnetic field tensor which is a rank 2 tensor. We need one direction to describe the electric field and another one to describe the magnetic field.

We can construct a tensor by using the matrix product ("row times column") of two vectors[22]

$$\vec{v}\vec{w}^T = \begin{pmatrix} v_1 \\ v_2 \\ v_3 \end{pmatrix} \begin{pmatrix} w_1 & w_2 & w_3 \end{pmatrix} = \begin{pmatrix} v_1 w_1 & v_1 w_2 & v_1 w_3 \\ v_2 w_1 & v_2 w_2 & v_2 w_3 \\ v_3 w_1 & v_3 w_2 & v_3 w_3 \end{pmatrix} \qquad (A.11)$$

[21] Using this definition, we can also understand scalars and vectors as tensors. A scalar encodes a magnitude and zero directions. For this reason, a scalar is often called a rank 0 tensor. A vector encodes a magnitude and exactly one direction. Therefore, we can call a vector a rank 1 tensor.

[22] Take note that this is not the dot product, which is defined as

$$\vec{v} \cdot \vec{w} = \vec{v}^T \vec{w} = \begin{pmatrix} v_1 & v_2 & v_3 \end{pmatrix} \begin{pmatrix} w_1 \\ w_2 \\ w_3 \end{pmatrix}$$

and yields a simple number. The superscript "T" denotes transposition.

A useful way to think about scalars, vectors and tensors is in terms of how they react to transformations of our coordinate system, e.g., rotations.

▷ A scalar remains completely unchanged.

▷ A vector transforms exactly like a position vector $\vec{r}$. For example, if we rotate our coordinate system using a rotation matrix R, i.e. $\vec{r} \to R\vec{r}$, any vector $\vec{v}$ gets rotated completely analogous to how $\vec{r}$ gets rotated: $\vec{v} \to R\vec{v}$.

▷ The transformation behavior of a tensor is more complicated. For example, a rank 2 tensor transforms like the product of two vectors in Eq. A.11: $M \to RMR^T$. We can understand this by transforming the product in Eq. A.11 explicitly:

$$M \equiv \vec{v}\vec{w}^T \to \left(R\vec{v}\right)\left(R\vec{w}\right)^T = R\vec{v}\vec{w}^T R^T \equiv RMR^T,$$

where we used that each vector transforms like a position vector, $\vec{v} \to R\vec{v}$, $\vec{w} \to R\vec{w}$. In general, we need one transformation matrix per tensor rank.[23]

[23] From this perspective it again makes sense to call a scalar a rank 0 tensor, since we need no transformation matrix.

The second statement may seem strange or even trivial. However, this definition is actually useful since, in principle, we can write any three quantities below each other between two big brackets. For example, we could write the pressure P, temperature T and entropy E of a gas between two big brackets

$$\begin{pmatrix} P \\ T \\ E \end{pmatrix}.$$

But still this object is not a vector since it doesn't transform like a position vector. These kind of thoughts are especially important in the context of special relativity. In special relativity our main focus are events in spacetime, which we can describe using four-vectors $x_\mu = (ct, x_1, x_2, x_3)^T$.[24]

An event is characterized by a location (x_1, x_2, x_3) and a point in time t. It makes sense to define four-vectors in special relativity since t and (x_1, x_2, x_3) are mixed through transformations

[24] Take note that the speed of light c appears here and in other four-vectors since all components of a vector must have the same units since otherwise we can't mix them. Since t has units [s], we multiply it by the only fundamental velocity that we have: c. The result ct has units [$\frac{m}{s}$ s]=[m] which is the same as the other components.

of our coordinate system ("boosts"). This is analogous to how (x_1, x_2, x_3) are mixed through rotations and hence we write them together as a vector. A crucial task is then to identify which quantities transform together like a four-vector. A famous example is the four-momentum $p_\mu = (E/c, p_1, p_2, p_3)$, where $\vec{p} = (p_1, p_2, p_3)$ is the ordinary momentum vector, E the energy and c denotes the speed of light. Another example is the electromagnetic potential $A_\mu = (\phi/c, A_1, A_2, A_3)$, where ϕ denotes the electric potential and $\vec{A} = (A_1, A_2, A_3)$ the magnetic vector potential.[25]

[25] Take note that there are objects which transform non-trivially but not like a position vector. The most famous example are spinors, which are mathematical objects that we need to describe elementary particles like electrons in quantum field theory.

In modern physics, simple scalars, vectors and tensors are often not enough to describe what is going on. Instead, we need scalar fields, vector fields, and tensor fields. We will talk about these mathematical tools in a moment.

However, first we will talk about the geometrical meaning of the dot and cross product since these are extremely important in electrodynamics.[26]

[26] We will not discuss the outer product, since we don't need it in electrodynamics.

A.2 The dot product

As mentioned above, the dot product allows us to combine two vectors $\vec{v}, \vec{w}$ in such a way that the result is a number[27]

[27] If we combine two vector functions, we get a scalar function.

$$\vec{v} \cdot \vec{w} = \begin{pmatrix} v_1 \\ v_2 \\ v_3 \end{pmatrix} \cdot \begin{pmatrix} w_1 \\ w_2 \\ w_3 \end{pmatrix} = v_1 w_1 + v_2 w_2 + v_3 w_3 \,. \quad (A.12)$$

In words, we can summarize the idea behind it as follows:

The scalar product of two vectors $\vec{v} \cdot \vec{w}$ yields the projection of the first vector $\vec{v}$ onto the axis defined by the second vector $\vec{w}$ times the length of the second vector.

How does this interpretation fit together with the formula given in Eq. A.12?

[28] The easiest way to understand projections in general is to consider projections onto the coordinate axis $\vec{e}_x, \vec{e}_y, \vec{e}_z$. The projection of some vector $\vec{v}$ onto a coordinate axis like $\vec{e}_x$ is simply what we usually call the first component v_1. In words, the meaning of this component is how much our vector $\vec{v}$ spreads out in the x-direction. Analogously, the projection of $\vec{v}$ onto $\vec{e}_y$ is what we call the second component v_2 and it tells us how much $\vec{v}$ spreads out in the y-direction.

To understand this, we need to talk about the projection of some vector $\vec{v}$ onto the axis defined by a second vector $\vec{w}$.[28]

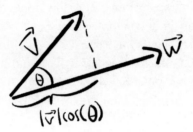

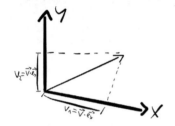

This allows us to write any vector in terms of basis vectors as follows

$$\vec{v} = v_1 \vec{e}_x + v_2 \vec{e}_y + v_3 \vec{e}_z. \quad (A.13)$$

By looking at the figure above, we can see that the correct formula for the projection of $\vec{v}$ onto $\vec{w}$ is

$$\text{projection of } |\vec{v}| \text{ onto the axis defined by } \vec{w} = |\vec{v}| \cos \theta \quad (A.14)$$

where θ denotes the angle between the two vectors. The statement from above in mathematical form therefore reads

$$\vec{v} \cdot \vec{w} = |\vec{v}| \cos \theta |\vec{w}|. \quad (A.15)$$

Therefore, the question we now need to answer is: how is this formula related to the usual formula in Eq. A.12?

To answer this question, we write our two general vectors in terms of our basis vectors:

$$\vec{v} = v_x \vec{e}_x + v_y \vec{e}_y + v_z \vec{e}_z$$
$$\vec{w} = w_x \vec{e}_x + w_y \vec{e}_y + w_z \vec{e}_z.$$

We can then rewrite our dot product in terms of dot products of the basis vectors:

$$\vec{v} \cdot \vec{w} = |\vec{v}||\vec{w}| \cos(\theta)$$
$$= \left[v_x \vec{e}_x + v_y \vec{e}_y + v_z \vec{e}_z \right] \cdot \left[w_x \vec{e}_x + w_y \vec{e}_y + w_z \vec{e}_z \right]$$
$$= v_x w_x (\vec{e}_x \cdot \vec{e}_x) + v_x w_y (\vec{e}_x \cdot \vec{e}_y) + v_x w_z (\vec{e}_x \cdot \vec{e}_z)$$
$$+ v_y w_x (\vec{e}_y \cdot \vec{e}_x) + v_y w_y (\vec{e}_y \cdot \vec{e}_y) + v_y w_z (\vec{e}_y \cdot \vec{e}_z)$$
$$+ v_z w_x (\vec{e}_z \cdot \vec{e}_x) + v_z w_y (\vec{e}_z \cdot \vec{e}_y) + v_z w_z (\vec{e}_z \cdot \vec{e}_z).$$

Next we use that our basis vectors are normalized ($\vec{e}_x \cdot \vec{e}_x = 1$) and orthogonal ($\vec{e}_x \cdot \vec{e}_y = 0$):

$$\begin{aligned}\vec{v} \cdot \vec{w} &= v_x w_x(1) + v_x w_y(0) + v_x w_z(0) \\ &+ v_y w_x(0) + v_y w_y(1) + v_y w_z(0) \\ &+ v_z w_x(0) + v_z w_y(0) + v_z w_z(1) \\ &= v_x w_x + v_y w_y + v_z w_z \quad \checkmark\end{aligned}$$

We can visualize this calculation as follows:

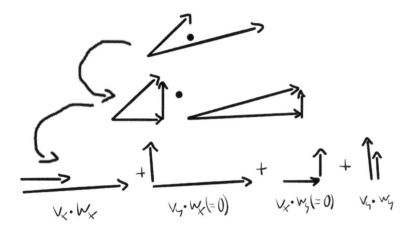

This tells us that we really can understand the result of the dot product as the projection of $\vec{v}$ onto $\vec{w}$ times the length of $\vec{w}$.

An important example is the dot product of a vector with itself, $\vec{v} \cdot \vec{v}$. In words, the result is the projection of $\vec{v}$ onto itself times the length of $\vec{v}$. Since the projection of $\vec{v}$ onto itself simply yields the full vector length, we simply get the length of the vector squared

$$\vec{v} \cdot \vec{v} = |\vec{v}| \cos 0 |\vec{v}| = |\vec{v}|^2 . \qquad (A.16)$$

> **Example: dot product of two vectors**
>
> The dot product of
> $$\vec{v} = \begin{pmatrix} 1 \\ 4 \\ 9 \end{pmatrix} \quad \text{and} \quad \vec{w} = \begin{pmatrix} 2 \\ 2 \\ 1 \end{pmatrix} \quad (A.17)$$
> is given by
> $$\vec{v} \cdot \vec{w} = \begin{pmatrix} 1 \\ 4 \\ 9 \end{pmatrix} \cdot \begin{pmatrix} 2 \\ 2 \\ 1 \end{pmatrix} = 2 + 8 + 9 = 19. \quad (A.18)$$

A.3 The cross product

As mentioned above, the cross product allows us to combine two vectors $\vec{A}, \vec{B}$ in such a way that the result is a vector[29]

[29] If we combine two vector functions, we get another vector function.

$$\vec{A} \times \vec{B} = \begin{pmatrix} A_1 \\ A_2 \\ A_3 \end{pmatrix} \times \begin{pmatrix} B_1 \\ B_2 \\ B_3 \end{pmatrix} = \begin{pmatrix} A_2 B_3 - A_3 B_2 \\ A_3 B_1 - A_1 B_3 \\ A_1 B_2 - A_2 B_1 \end{pmatrix}. \quad (A.19)$$

In words, we can summarize the idea behind it as follows:[30]

[30] The direction in which the resulting vector points can be determined by the right-hand rule.

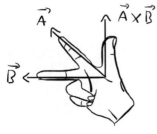

> The cross product of two vectors $\vec{A} \times \vec{B}$ yields a vector perpendicular to $\vec{A}$ and $\vec{B}$ whose magnitude is the area of the parallelogram spanned by $\vec{A}$ and $\vec{B}$.

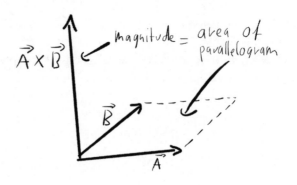

Now, how does this interpretation fit together with the formula in Eq. A.19?

To understand this, we first need to recall that the formula for the area of a parallelogram is base times height. Here, our base is given by the length of the vector $\vec{A}$ and the height by $\sin(\theta)|\vec{B}|$, where θ is the angle between $\vec{A}$ and $\vec{B}$.

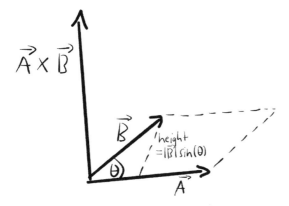

The area of the parallelogram spanned by $\vec{A}$ and $\vec{B}$ is therefore

$$\text{area} = |\vec{A}| \sin(\theta) |\vec{B}|. \tag{A.20}$$

Therefore, the question we now need to answer is: how is this formula related to the usual formula in Eq. A.19?

To answer this question, we write our two general vectors in terms of our basis vectors:

$$\vec{v} = v_x \vec{e}_x + v_y \vec{e}_y + v_z \vec{e}_z$$
$$\vec{w} = w_x \vec{e}_x + w_y \vec{e}_y + w_z \vec{e}_z.$$

We can then rewrite our cross product in terms of cross products of the basis vectors:

$$|\vec{v} \times \vec{w}| = \left|[v_x\vec{e}_x + v_y\vec{e}_y + v_z\vec{e}_z] \times [w_x\vec{e}_x + w_y\vec{e}_y + w_z\vec{e}_z]\right|$$

$$= |v_xw_x(\vec{e}_x \times \vec{e}_x) + v_xw_y(\vec{e}_x \times \vec{e}_y) + v_xw_z(\vec{e}_x \times \vec{e}_z)$$
$$+ v_yw_x(\vec{e}_y \times \vec{e}_x) + v_yw_y(\vec{e}_y \times \vec{e}_y) + v_yw_z(\vec{e}_y \times \vec{e}_z)$$
$$+ v_zw_x(\vec{e}_z \times \vec{e}_x) + v_zw_y(\vec{e}_z \times \vec{e}_y) + v_zw_z(\vec{e}_z \times \vec{e}_z)|.$$

Next we use that the cross product of a vector with itself yields zero ($\vec{e}_x \times \vec{e}_x = 0$) and that the cross product of two basis vectors yields the third basis vector ($\vec{e}_x \times \vec{e}_y = \vec{e}_z$):[31]

$$|\vec{v} \times \vec{w}| = |v_xw_x(0) + v_xw_y(\vec{e}_z) + v_xw_z(-\vec{e}_y)$$
$$+ v_yw_x(-\vec{e}_z) + v_yw_y(0) + v_yw_z(\vec{e}_x)$$
$$v_zw_x(\vec{e}_y) + v_zw_y(-\vec{e}_x) + v_zw_z(0)|$$
$$= |(v_yw_z - v_zw_y)\vec{e}_x + (v_zw_x - v_xw_z)\vec{e}_y + (v_xw_y - v_yw_x)\vec{e}_z|$$

[31] This is necessarily the case since the vector that we find by calculating the cross product of two vectors, $\vec{a} \equiv \vec{b} \times \vec{c}$, is orthogonal to the two vectors in the product. Here $\vec{a}$ is orthogonal to $\vec{b}$ and $\vec{c}$.

Therefore, we can conclude that the cross product really yields a vector whose length is the area of the parallelogram spanned by $\vec{A}$ and $\vec{B}$.

Example: cross product of two vectors

The cross product of

$$\vec{A} = \begin{pmatrix} 1 \\ 4 \\ 9 \end{pmatrix} \quad \text{and} \quad \vec{B} = \begin{pmatrix} 2 \\ 2 \\ 1 \end{pmatrix} \tag{A.21}$$

is given by

$$\vec{A} \times \vec{B} = \begin{pmatrix} 1 \\ 4 \\ 9 \end{pmatrix} \times \begin{pmatrix} 2 \\ 2 \\ 1 \end{pmatrix} = \begin{pmatrix} 4 - 18 \\ 18 - 1 \\ 2 - 8 \end{pmatrix} = \begin{pmatrix} -14 \\ 17 \\ -6 \end{pmatrix}. \tag{A.22}$$

A.4 Fields

An ordinary function $f(x)$ is an object which eats some number x and spits out another number $f(x)$, e.g., $f(3) = 7$. In physics, we often need functions which eat a location $\vec{r}$ and spit out a number $f(\vec{r})$, e.g., $f(2,3,1) = 9$. We usually call an object like this a **scalar field**.[32]

[32] In physics, a field is a quantity which exists everywhere in space at the same time. This is in contrast to ordinary objects like, say, a ball which only exists at one particular location.

A scalar field assigns a number to each point in space.[33] This number is what we call the **field strength** at the given point.

[33] Reminder: scalar is another word for a simple number. The defining feature of a scalar is that it remains completely unchanged under coordinate transformations like, for example, rotations. In contrast, a vector, in general, gets changed if we rotate our coordinate system.

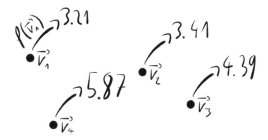

A scalar field is the proper mathematical tool to describe, for example, temperature. The value of the temperature field at each point is simply the temperature there.

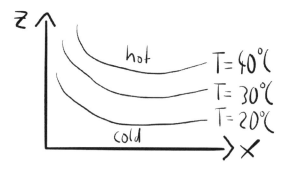

Completely analogously, we can define so-called **vector fields**. While a scalar field assigns a simple number to each point in space, a vector field assigns vectors. The length of the vector at each location represents the field strength at the point. In addition, the vector also defines a direction.

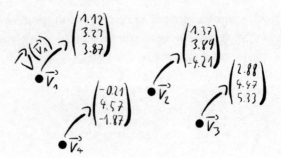

A vector field is the proper mathematical tool to describe, for example, air. The field strength (vector length) at each point represents the velocity of the air molecules. Moreover, the direction in which the vector points at each location encodes in which direction the air flows.

Analogously, we can also define so-called **tensor fields**. A tensor field assigns a tensor to each point in space.[34]

[34] Reminder: a rank 2 tensor is a matrix. Higher rank tensors are even more abstract objects. A scalar is simply a number and can be treated as a rank 0 tensor. A vector has exactly one index and can therefore be treated as a rank 1 tensor.

A tensor field is useful, for example, to describe inertia. The matrix at each point in this case encodes information about how difficult it is to move in each of the three directions (x, y, z).[35]

[35] Another way to think about it is as three vectors attached to each point.

To summarize, we use scalar fields to describe quantities which are completely described by their magnitude at each point in space. In contrast, we use vector fields to describe quantities which are completely described by their magnitude and a direction at each point in space. Tensor fields are useful to describe quantities for which we need more than just a magnitude and a direction to describe them.

Formulated differently:

▷ A scalar field eats a location $\vec{r}_1 = (1,3,4)^T$ and spits out a number, e.g., $\phi(\vec{r}_1) = 3$. For a different location $\vec{r}_2 = (9,2,4)^T$, we possibly get a different number: $\phi(\vec{r}_2) = 9.2$.

▷ A vector field eats a location $\vec{r}_1 = (1,3,4)$ and spits our a vector, e.g., $\vec{A}(\vec{r}_1) = (1,3,2)^T$. Again, if we put in a different location $\vec{r}_2 = (9,2,4)^T$, we possibly get a different vector: $\vec{A}(\vec{r}_1) = (3,11,9)^T$.

▷ A tensor field eats a location $\vec{r}_1 = (1,3,4)^T$ and spits out a tensor, e.g., $M(\vec{r}_1) = \begin{pmatrix} 1 & 2 & 3 \\ 3 & 4 & 5 \\ 3 & 4 & 5 \end{pmatrix}$. If we put in a different location $\vec{r}_2 = (9,2,4)^T$, we possibly get a different tensor: $M(\vec{r}_2) = \begin{pmatrix} 8 & 9 & 9 \\ 4 & 3 & 22 \\ 3 & 3 & 15 \end{pmatrix}$.

Take note that in the following, we will use the notions of "scalar field" and "scalar function" or "vector field" and "vector function" interchangeably.

A.5 Line integral

The simplest kind of integral involving fields is when we integrate a scalar field ϕ over a curve C:

$$\int_C \phi(\vec{r}) dl \, .$$

We call an integral of this form a **line integral** and its meaning can be summarized as follows

> We calculate the magnitude of the scalar field $\phi(\vec{r})$ at each point on the curve C and then sum over all these individual contributions.

This type of integral is useful, for example, to calculate the total mass of a thin wire if only the (non-constant) mass density $\rho(x)$ is known or to calculate the total charge if only the charge density is known.

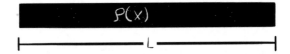

Let's discuss the idea behind the line integral in a bit more detail.

The main idea is that we divide the curve C, e.g. our wire, into N short segments Δx_i.

We can then calculate, for example, the mass contained in each short segment by multiplying the length of the segment Δx_i with the mass density ρ_i in this region:

$$\text{mass in each segment i} = \rho_i \Delta x_i . \tag{A.23}$$

The total mass of the wire is then given by the sum over these N individual contributions

$$\text{mass of the wire} \simeq \sum_i^N \rho_i \Delta x_i . \tag{A.24}$$

This formula is not quite exact since we need to use some average value of the mass density ρ_i in each segment. The formula becomes exact in the limit where the length of each segment Δx_i goes to zero. In this limit, the sum becomes an integral and we are then left with

$$\text{mass of the wire} = \int_0^L \rho(x) dx, \tag{A.25}$$

where L denotes the length of the wire.

Take note that it's also possible to consider more complicated curves C. In general, we have to find a parametrization of the curve $\vec{r}(l)$, where l is a parameter like, for example, time. In the simple case above, our curve was a line and we can parameterize it using $\vec{r} = (x, 0, 0)^T$ with x running from 0 to L.

Example: line integral over circular path

If we consider, for example, a circular path we have to use a parameterization of the form

$$\vec{r}(\varphi) = \begin{pmatrix} R\cos(\varphi) \\ R\sin(\varphi) \\ 0 \end{pmatrix}, \quad (A.26)$$

where R denotes the radius of the circle.

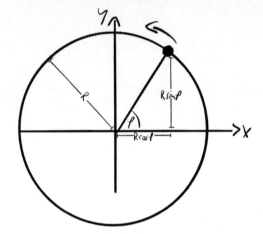

The line integral of a specific scalar function, for example,

$$\phi(\vec{x}) = x^2 y^2,$$

can then be calculated as follows

$$\begin{aligned}
\oint_C \phi(\vec{r})\, dl &= \int_0^{2\pi} \phi(\vec{r})\, R\, d\varphi \\
&= \int_0^{2\pi} (R\cos(\varphi))^2 (R\sin(\varphi))^2\, R\, d\varphi \\
&= R^5 \frac{\pi}{4}.
\end{aligned} \quad (A.27)$$

Take note that the symbol $\oint$ is used to indicate an integral over a closed curve.

A.6 Path integral

The **path integral** is defined as the integral of a vector field $\vec{V}(\vec{x})$ over some path P:

$$\int_P \vec{V} \cdot d\vec{l} = \int_P \vec{V} \cdot \vec{t}(l) dl \, ,$$

where l parameterizes the path P and $\vec{t}(l)$ yields a vector tangential to the path at each point l on the path.

In words, the path integral can be described as follows[36]

> We calculate the component of the vector field $\vec{V}$ in the direction $d\vec{l}$ of the path P at each point. The path integral is then the sum over all these individual contributions.

[36] Reminder: as discussed in Appendix A.2, the dot product yields the projection of the first vector onto the direction defined by the second vector.

The path integral is important, for example, to calculate the work done by some force and also appears in Maxwell's equations.

Let's discuss how this interpretation of the path integral comes about in detail.

Using the line integral discussed in the previous section, we calculate a sum over the values of a scalar function $\phi(\vec{x})$ on some line C. In other words, we calculate a sum over the magnitude of $\phi(\vec{x})$ in some region defined by the line L.

Now, if we want to integrate a vector function $\vec{A}(\vec{x})$, there is one additional thing that we need to take into account. The main difference between a scalar function and a vector function is that the latter not only has a magnitude at each point in space, but also a direction.[37] Therefore, when we integrate over a

[37] The difference between scalar and vector functions is discussed in Appendix A.4.

vector function $\vec{A}(\vec{x})$, we somehow need to take this additional information into account.

To understand this, let's consider a concrete example.

The increment of work W done by a specific force $\vec{F}$ is given by the product of the magnitude of the force times the displacement of the object Δx:

$$\Delta W = |\vec{F}| \Delta x. \tag{A.28}$$

However, this simple formula is only correct if the force is constant and if the displacement is in the same direction as the force.[38] We therefore need to refine our formula to take these two possibilities into account.

[38] To understand how the displacement can be in a different direction than the force, imagine an object flying with some initial velocity. While the force will push the object in a specific direction, the object will still fly forward. The total displacement is then a sum of the displacement caused by the force and its forward movement.

First of all, if the force is not constant, we can use the same trick that we already used for the line integral: we divide our path P into N short segments Δl_i and then sum over all these individual contributions

$$W = \sum_i^N \Delta W_i \simeq \sum_i^N |\vec{F}_i| \Delta l_i, \tag{A.29}$$

where $\vec{F}_i$ denotes the force applied in the segment Δl_i. In the limit of vanishing segment length, we again end up with an integral[39]

$$W = \int_P |\vec{F}(l)| \, dl, \tag{A.30}$$

[39] Since $|\vec{F}(l)|$ is a scalar function, we have a simple line integral.

where l parameterizes our path P.

Now, how can we take into account that, in general, the displacement is not in the same direction as the force? The crucial idea is that we only count the component of the force in the direction of the displacement $\Delta \vec{l}_i$. In other words, we need to project the force onto our path P and only then sum over these contributions. Luckily, we already know the mathematical tool which allows us to this: the dot product.[40] The component of the force in the direction of the displacement $\Delta \vec{l}_i$ is

[40] The dot product is discussed in Appendix A.2.

$$\Delta W = \vec{F} \cdot \Delta \vec{l}_i = |\vec{F}||\Delta \vec{l}_i| \cos(\theta), \tag{A.31}$$

where θ is the angle between $\vec{F}$ and $\Delta \vec{l}_i$.

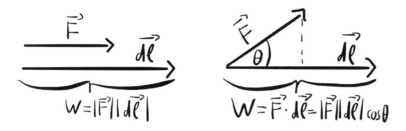

Therefore, in the most general case, we need to calculate $\vec{F} \cdot \Delta \vec{l}_i$ for each segment and then sum over these contributions

$$W = \int_P \vec{F}(\vec{l}) \cdot d\vec{l}, \tag{A.32}$$

This type of integral is known as a path integral.

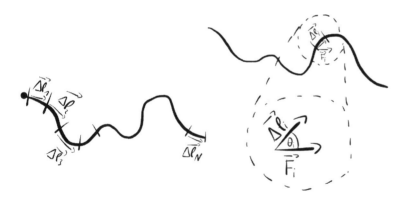

The path integral over a circular path $\oint_P \vec{V}(\vec{l}) \cdot d\vec{l}$ is an extremely important special case which shows up, for example, in Maxwell's equations. The resulting quantity that we get when we integrate a vector field over a circular path is known as the circulation of the vector field.[41]

[41] The circulation of a vector field is the topic of Appendix A.7.

Now, a question we still need to answer is: How can we calculate $d\vec{l}$ for complicated paths? This is the topic of the next section.

A.6.1 Tangent vector

As mentioned in the previous section, the vector $d\vec{l}$ at each point on the path P points in the direction of the displacement.

For this reason, $d\vec{l}$ is known as a **tangent vector**. At each point in space, we have a different tangent vector and for this reason it is often helpful to write

$$d\vec{l} = \vec{t}(s)ds, \qquad (A.33)$$

where s parameterizes our path P.

Now how can we calculate $\vec{t}(s)$? To understand this, imagine an object moves along the path P. We describe the path using a specific trajectory $\vec{r}(s)$, where you can think of s, for example, as the time. Our task is to find a vector which at each point s is tangential to the path $\vec{r}(s)$.

Luckily, in physics there is one extremely familiar vector function with exactly this property: the velocity of the object. The velocity encodes the rate at which the location of our object changes and, in addition, also in which direction it moves. In particular, given the location of our object at some specific point in time s_0, we can calculate the location at a later point in time s_1 using

$$\vec{r}(s_1) \simeq \vec{r}(s_0) + \vec{v}(s_0)\Delta s \qquad (A.34)$$

where $\Delta s \equiv s_1 - s_0$.

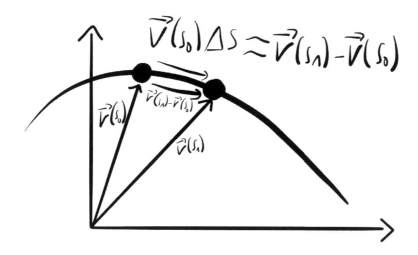

This means that the velocity vector function

$$\vec{v}(s) \equiv \frac{d\vec{r}(s)}{ds} \tag{A.35}$$

yields at each point s a vector tangential to the path $\vec{r}(s)$.

To get the normalized tangent vectors, all we have to do is divide $\vec{v}(s)$ by its length[42]

$$\vec{t}(s) = \frac{\vec{v}(s)}{|\vec{v}(s)|}. \tag{A.36}$$

This is how we can calculate the tangent vectors for a general path P.

We can also see this formally as follows

$$\int \vec{V} \cdot d\vec{r} = \int \vec{V} \cdot \frac{d\vec{r}}{dt} dt = \int \vec{V} \cdot \vec{v} dt \tag{A.37}$$

[42] We use normalized tangent vectors since, for example, the work done often does not depend on the velocity. All we need is a tangent vector which allows us to project the vector field onto the path. An additional factor, like the velocity, can always be added additionally, if necessary; but is not essential to the definition of a path integral.

Example: work done along a rectangular path

For concreteness, let's imagine we have a force described by the vector field

$$\vec{F} = \begin{pmatrix} Lx^2 - y^3 \\ -xy^2 \\ xyz \end{pmatrix} \tag{A.38}$$

and we want to calculate the work done if we apply this force along a small square with edge length $2L$ around the origin in the xy plane.

252 NO-NONSENSE ELECTRODYNAMICS

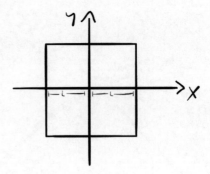

To calculate the path integral

$$\oint \vec{F} \cdot d\vec{r} = \oint \begin{pmatrix} Lx^2 - y^3 \\ -xy^2 \\ xyz \end{pmatrix} \cdot d\vec{r}, \quad (A.39)$$

we divide the square into four parts.

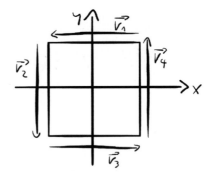

These can be parameterized by

1. $\vec{r}_1(t) = \begin{pmatrix} x \\ y \\ z \end{pmatrix} = \begin{pmatrix} t \\ L \\ 0 \end{pmatrix}, \frac{d\vec{r}_1}{dt} = \begin{pmatrix} 1 \\ 0 \\ 0 \end{pmatrix}$, where t runs from L to $-L$.

2. $\vec{r}_2(t) = \begin{pmatrix} x \\ y \\ z \end{pmatrix} = \begin{pmatrix} -L \\ t \\ 0 \end{pmatrix}, \frac{d\vec{r}_2}{dt} = \begin{pmatrix} 0 \\ 1 \\ 0 \end{pmatrix}$, where t runs from L to $-L$.

3. $\vec{r}_3(t) = \begin{pmatrix} x \\ y \\ z \end{pmatrix} = \begin{pmatrix} t \\ -L \\ 0 \end{pmatrix}, \frac{d\vec{r}_3}{dt} = \begin{pmatrix} 1 \\ 0 \\ 0 \end{pmatrix}$, where t runs from $-L$ to L.

4. $\vec{r}_4(t) = \begin{pmatrix} x \\ y \\ z \end{pmatrix} = \begin{pmatrix} L \\ t \\ 0 \end{pmatrix}, \frac{d\vec{r}_4}{dt} = \begin{pmatrix} 0 \\ 1 \\ 0 \end{pmatrix}$, where t runs from $-L$ to L.

The total path integral is therefore

$$\oint_\square \begin{pmatrix} Lx^2 - y^3 \\ -xy^2 \\ 0 \end{pmatrix} \cdot d\vec{r} = \oint_\square \begin{pmatrix} Lx^2 - y^3 \\ -xy^2 \\ 0 \end{pmatrix} \cdot \frac{d\vec{r}}{dt} dt$$

$$= \int_L^{-L} \begin{pmatrix} Lt^2 - L^3 \\ -tL^2 \\ 0 \end{pmatrix} \cdot \begin{pmatrix} 1 \\ 0 \\ 0 \end{pmatrix} dt$$

$$+ \int_L^{-L} \begin{pmatrix} L^3 - t^3 \\ Lt^2 \\ 0 \end{pmatrix} \cdot \begin{pmatrix} 0 \\ 1 \\ 0 \end{pmatrix} dt$$

$$+ \int_{-L}^{L} \begin{pmatrix} Lt^2 + L^3 \\ -tL^2 \\ 0 \end{pmatrix} \cdot \begin{pmatrix} 1 \\ 0 \\ 0 \end{pmatrix} dt$$

$$+ \int_{-L}^{L} \begin{pmatrix} L^3 - t^3 \\ -Lt^2 \\ 0 \end{pmatrix} \cdot \begin{pmatrix} 0 \\ 1 \\ 0 \end{pmatrix} dt$$

$$= -\int_{-L}^{L} (Lt^2 - L^3) dt - \int_{-L}^{L} Lt^2 dt$$
$$+ \int_{-L}^{L} (Lt^2 + L^3) dt + \int_{-L}^{L} (-Lt^2) dt$$

$$= \int_{-L}^{L} (2L^3 - 2Lt^2) dt = [2L^3 t - \frac{2}{3} Lt^3]_{-L}^{L}$$

$$= 4L^4 - \frac{4}{3} L^4 = \frac{8}{3} L^4. \tag{A.40}$$

A.7 Circulation integral

The circulation of a vector field $\vec{V}(\vec{x})$ is defined as

$$\boxed{\text{Circulation}(\vec{x}) = \oint_P \vec{V}(\vec{x}) \cdot \vec{t} dt,}$$

where t is a parameter which parameterizes our path and $\vec{t}$ a unit vector tangential to the path. In other words, the circulation is simply the path integral over the vector field for a circular path.[43]

Intuitively, as the name already indicates, it's a measure for how much our vector field circulates.

[43] Information about the circulation at a single point is encoded in the curl of the vector field. This is the topic of Appendix A.12.

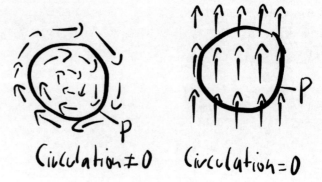

A.8 Surface integral

Given a scalar field $\phi(\vec{x})$, we can not only calculate the integral over some curve C, but also the integral over a surface S:

$$\int_S \phi(\vec{x}) da$$

> We calculate the magnitude of the scalar field $\phi(\vec{x})$ at each point on the surface S and then sum over all these individual contributions.

This type of integral is useful, for example, to calculate the total mass of a thin plate if only the (non-constant) mass density $\rho(\vec{x})$ is known or to calculate the total charge if only the charge density is known.

Let's discuss how this interpretation of the surface integral comes about in detail.

The main idea is completely analogous to what we already discussed for the line integral in Appendix A.5. We divide our surface S into small segments δa and then use that the mass of each segment is given by the product of δa and the mass surface density in this region ρ_i:

$$\text{mass in each segment i} = \rho_i \Delta a_i. \qquad (A.41)$$

The total mass of the plate is then given by the sum over these N individual contributions

$$\text{mass of the plate} \simeq \sum_i^N \rho_i \Delta a_i. \qquad (A.42)$$

This formula is not quite exact since we need to use some average value of the mass density ρ_i in each segment. The formula becomes exact in the limit where the area of each segment Δa_i goes to zero. In this limit, the sum becomes an integral and we are then left with

$$\text{mass of the plate} = \int_S \rho(\vec{x}) da. \qquad (A.43)$$

A.8.1 Example: surface integral

Let's imagine, we want to calculate the surface integral for the scalar function $\phi(\vec{x}) = x^2 y^2$ over a square around the origin in the xy-plane.

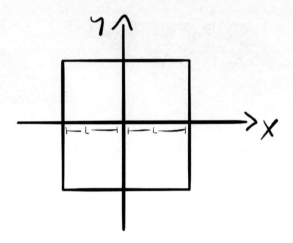

The surface integral can therefore be calculated as follows

$$
\begin{aligned}
\int_\square \phi(\vec{x}) da &= \int_{-L}^{L} \int_{-L}^{L} \phi(\vec{x})\, dxdy \\
&= \int_{-L}^{L} \int_{-L}^{L} x^2 y^2 \, dxdy \\
&= \left[\int_{-L}^{L} \frac{1}{3} x^3 y^2 \, dy \right]_{-L}^{L} \\
&= \int_{-L}^{L} \frac{1}{3} L^3 y^2 \, dy - \int_{-L}^{L} \frac{1}{3} (-L)^3 y^2 \, dy \\
&= \frac{2L^3}{3} \int_{-L}^{L} y^2 \, dy \\
&= \frac{2L^3}{3} \left[\frac{1}{3} y^3 \right]_{-L}^{L} \\
&= \frac{2L^3}{3} \left(\frac{1}{3} L^3 - \frac{1}{3} (-L)^3 \right) \\
&= \frac{4L^6}{9}.
\end{aligned}
\tag{A.44}
$$

A.9 Flux Integral

Given a vector field $\vec{V}(\vec{x})$, we can not only calculate the integral *along* some path P, but also the integral *through* a surface S:[44]

$$\int_S \vec{V} \cdot d\vec{a} = \int_S \vec{V} \cdot \vec{n} da ,$$

where $\vec{n}$ is a unit vector normal to the surface S. In words, the surface integral can be described as follows[45]

> We calculate the component of the vector field $\vec{V}$ normal to the surface at each point on the surface. The surface integral is then the sum over all these individual contributions.

The flux integral is important, for example, to describe how a fluid flows through a pipe and, of course, because it appears in Maxwell's equations.

Let's discuss how this interpretation of the flux integral comes about in detail.

[44] We discussed the path integral in Appendix A.6.

[45] Reminder: as discussed in Appendix A.2, the dot product yields the projection of the first vector onto the direction defined by the second vector.

Using the surface integral discussed in the previous section, we calculate a sum over the values of a scalar function $\phi(\vec{x})$ on some surface S. In other words, we calculate a sum over the magnitude of $\phi(\vec{x})$ in some region S.

Now, if we want to integrate a vector function $\vec{V}(\vec{x})$ over some surface, there is again one additional thing that we need to take into account: a vector function also describes a particular direction at each point in space, not only a magnitude.[46] Therefore, when we integrate over a vector function $\vec{V}(\vec{x})$, we somehow need to take this additional information into account.

To understand this, let's consider a concrete example.

[46] The difference between scalar and vector functions is discussed in Appendix A.4.

[47] The current density describes in which direction and how much electric charge passes per unit area. We discuss the current density in detail in Section 2.3.

[48] The number density describes the number of particles per cubic meter.

Let's assume our vector field describes a current density $\vec{J}(\vec{x})$.[47] For concreteness, let's assume our current density is given by the number density $\rho(\vec{x})$ times their average velocity $\vec{v}$:[48]

$$\vec{J}(\vec{x}) = \rho(\vec{x})\vec{v}. \tag{A.45}$$

This quantity has units of particles per square meter per second. In words, it tells us how many particles pass each unit area per unit time. Therefore, if our current density is uniform $\vec{J}$ and perpendicular to some specific surface S, the flux through S is simply the

$$\text{number of particles per second through } S \equiv \text{flux through } S$$
$$\equiv |\vec{J}| \times \text{ surface area}.$$

However, analogous to what we did in Appendix A.6, this formula needs to be refined if we want to calculate the flux in more general situations.

If the current density is not constant, we can use the same trick that we already used for the surface integral: we divide our surface S into N small segments Δa_i and then sum over all these individual contributions

$$\text{flux through } S \simeq \sum_{i}^{N} |\vec{J}_i| \Delta a_i, \tag{A.46}$$

where $\vec{J}_i$ is the average current density in the region Δa_i. In the limit of vanishing segment areas, we end up again with an integral[49]

[49] Since $|\vec{J}(\vec{x})|$ is a scalar function, we simply have a surface integral.

$$\text{flux through } S = \int_S |\vec{J}| da. \tag{A.47}$$

Now, how can we take into account the direction of our vector field at each point on the surface?

In our concrete example, the direction of the vector at each point indicates the direction in which our particles flow. Therefore, if we want to calculate how many particles really pass a specific surface, we need to be careful since, in general, the full flow does not "hit" the surface.

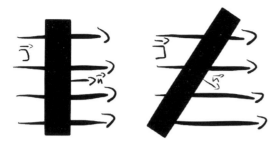

This means that if we use Eq. A.47, we would make a big mistake since far fewer particles really pass the surface S. To understand this problem, we use the fact that we can split up our general vector $\vec{J}(\vec{r}_0)$ at a particular point $\vec{r}_0$ on the surface in terms of a component tangential $J_\parallel$ to the surface and a component $J_\perp$ normal to the surface

$$\vec{J}(\vec{r}_0) = J_\perp \vec{e}_\perp + J_\parallel \vec{e}_\parallel . \qquad (A.48)$$

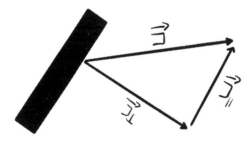

This is useful since the component tangential to the surface can be neglected completely. A current density vector tangential to the surface means that the particles flow along the surface and therefore do not penetrate it.

In general, we can find the relevant component of $\vec{J}(\vec{x})$ at each point on the surface by using the dot produce once more. In particular, the dot product between $\vec{J}(\vec{x})$ and a unit vector $\vec{n}$ normal to the surface S yields exactly the relevant component $J_\perp$.[50]

[50] Recall that the dot product between a vector and a unit vector yields exactly the component of vector along the direction of the unit vector.

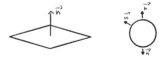

Therefore, in the most general case, we need to calculate $\vec{J} \cdot \Delta \vec{a}_i$ for each segment and then sum over these contributions[51]

$$\text{flux through } S = \int_S \vec{J}(\vec{x}) \cdot d\vec{a} = \int_S \vec{J}(\vec{x}) \cdot \vec{n} da. \qquad (A.49)$$

Again an important special case is the flux through a *closed* surface. In this case, we usually denote the flux like this

$$\text{flux through closed surface } S = \oint_S \vec{J}(\vec{x}) \cdot d\vec{a} = \oint_S \vec{J}(\vec{x}) \cdot \vec{n} da. \qquad (A.50)$$

[51] Take note that, in general, we have a different normal vector at different points on the surface.

Example: flux integral

Let's imagine that we want to calculate the flux integral of the vector function

$$\vec{F} = \begin{pmatrix} xz \\ -yz \\ 2y^2 \end{pmatrix} \tag{A.51}$$

over a square around the origin in the xy-plane.

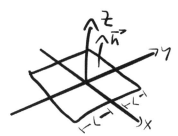

The unit vector normal to the square is

$$\vec{n} = \begin{pmatrix} 0 \\ 0 \\ 1 \end{pmatrix}. \tag{A.52}$$

The flux integral can therefore be calculated as follows

$$\int_{\square} \vec{F} \cdot d\vec{a} = \int_{\square} \begin{pmatrix} xz \\ -yz \\ 2y^2 \end{pmatrix} \cdot \begin{pmatrix} 0 \\ 0 \\ 1 \end{pmatrix} da$$

$$= 2 \int_A y^2 da$$

$$= 2 \int_{-L}^{L} dx \int_{-L}^{L} y^2 dy$$

$$= 2[x]_{-L}^{L} [\frac{1}{3}y^3]_{-L}^{L}$$

$$= 2 \cdot 2L \cdot \frac{2}{3} L^3 = \frac{8}{3} L^4 \tag{A.53}$$

A.10 Gradient

The gradient is an operator which transforms a scalar function $f(\vec{x})$ into a vector function:[52]

[52] The symbol ∇ is usually called "del" or "nabla". It is usually written like a vector

$$\nabla = \begin{pmatrix} \partial_x \\ \partial_y \\ \partial_z \end{pmatrix}.$$

The nabla symbol is not only important here but is also used to get a scalar function from a given vector function using the divergence $\nabla \cdot \vec{v}(\vec{x})$, c.f. Appendix A.11. Moreover, we use the nabla symbol to define the curl of a vector field $\nabla \times \vec{v}(\vec{x})$, c.f. Appendix A.12.

$$\nabla f(\vec{x}) = \begin{pmatrix} \partial_x \\ \partial_y \\ \partial_z \end{pmatrix} f(\vec{x}) = \begin{pmatrix} \partial_x f(\vec{x}) \\ \partial_y f(\vec{x}) \\ \partial_z f(\vec{x}) \end{pmatrix}.$$

The meaning of the resulting vector function, called the gradient, can be summarized as follows:[53]

[53] We discussed flux in Appendix A.9.

> The gradient of a scalar function is a vector function $\vec{G}(\vec{x}) = \nabla f(\vec{x})$ which describes the rate of change of the scalar function in the three coordinate directions.

In particular, this means that the x-component $\partial_x f(\vec{x})$ of the resulting vector function $\nabla f(\vec{x})$ tells us the slope of the scalar field $f(\vec{x})$ in the x-direction. Analogously, the y-component $\partial_y f(\vec{x})$ tells us the slope in the y-direction and $\partial_z f(\vec{x})$ the slope in the z-direction. In this sense, the gradient is simply an extension of the usual derivative $\partial_x f(x)$ for functions which depend on more than one coordinate.

For example, let's imagine we have a scalar function $f(\vec{x})$ which describes the height of terrain above sea level. If we now evaluate the gradient $\partial_x f(\vec{x})$ at one specific location $\vec{G}(\vec{r})$ and use its first component, it tells us whether we move up or down if we move in the x-direction. In addition, it also tells us how steep the slope is. To describe the slope in any possible direction, we need the slope in the three basis directions x, y and z and this is why the gradient has three components.[54]

[54] Take note that the gradient assigns a vector to each location. This vector points uphill in the direction of steepest slope and its magnitude $|\nabla f(\vec{x})|$ yields the slope in this particular direction.

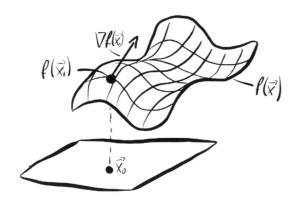

Example: gradient

The gradient of the scalar field

$$\phi(\vec{x}) = x^2 y + y^2 - zx \qquad (A.54)$$

is

$$\nabla \phi(\vec{x}) = \begin{pmatrix} \partial_x \\ \partial_y \\ \partial_z \end{pmatrix} \left(x^2 y + y^2 - zx \right)$$

$$= \begin{pmatrix} \partial_x \left(x^2 y + y^2 - zx \right) \\ \partial_y \left(x^2 y + y^2 - zx \right) \\ \partial_z \left(x^2 y + y^2 - zx \right) \end{pmatrix}$$

$$= \begin{pmatrix} 2xy - z \\ x^2 + 2y \\ -x \end{pmatrix}. \qquad (A.55)$$

A.11 Divergence

The divergence $\nabla \cdot$ is an operator which transforms a vector function $\vec{F}(\vec{x}) = (F_x(x), F_y(x), F_z(x))^T$ into a scalar function:[55]

[55] Reminder: the little dot $\cdot$ denotes the scalar product.

$$\nabla \cdot \vec{F}(\vec{x}) = \begin{pmatrix} \partial_x \\ \partial_y \\ \partial_z \end{pmatrix} \cdot \begin{pmatrix} F_x(\vec{x}) \\ F_y(\vec{x}) \\ F_z(\vec{x}) \end{pmatrix} = \partial_x F_x(\vec{x}) + \partial_y F_y(\vec{x}) + \partial_z F_z(\vec{x}).$$

The meaning of the resulting scalar function, called the divergence, can be summarized as follows:[56]

[56] We discussed flux in Appendix A.9.

> The divergence of a vector function is a scalar function $D(\vec{x}) = \nabla \cdot \vec{F}(\vec{x})$ which describes the amount of flux $D(\vec{x})$ entering or leaving a given point $\vec{x}$.

So, analogous to how the charge density $\rho(\vec{x})$ encodes information about how the total charge q is distributed throughout the system, the divergence $\nabla \cdot \vec{F}(\vec{x})$ tells us how flux is distributed. While the charge density $\rho(\vec{x})$ is defined as net charge per unit volume, the divergence $\nabla \cdot \vec{F}(\vec{x})$ is defined as net flux per unit volume:[57]

[57] We will discuss below how this interpretation fits together with the definition in terms of derivatives given above.

$$\text{Divergence} = \frac{\text{Flux}}{\text{Volume}}. \tag{A.56}$$

In this sense, an alternative name for the divergence of a vector field could be "flux density".

From a slightly different perspective, we can say that the divergence of a vector field tells us how much the field "diverges" from a point.[58]

[58] In physics, how much a field diverges from a point depends on how much charge is present at this point. We will talk about this in more detail below.

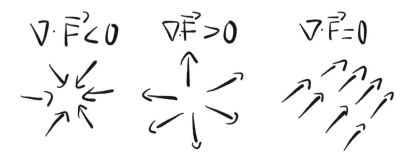

Let's talk about these definitions and interpretations in a bit more detail.

As mentioned above, our goal is to define a quantity which describes the flux of a vector field at a *single* point.[59]

In Section A.9, we defined the flux through a surface S as

$$\phi = \int_S \vec{F} \cdot d\vec{S}. \tag{A.57}$$

The key idea is now that we shrink the surface S until we are left with the neighborhood of a single point. However, if we simply take the limit $\lim_{S \to 0}$ in Eq. A.57 we find that the flux vanishes:

$$\lim_{S \to 0} \int_S \vec{F} \cdot d\vec{S} = 0. \tag{A.58}$$

Therefore, this is not a useful quantity to describe the flux at a single point since it is simply zero for every point and thus contains no specific information. To get a meaningful quantity, we need to somehow account for the vanishing surface area in such a way that the total result is something constant, not something vanishing.

A clever idea is to divide Eq. A.57 by the volume corresponding to the surface S. Each time we divide our surface area the volume gets divided too. Hence, the effect induced by the vanishing surface area is canceled by the vanishing volume and the

[59] The main motivation to introduce this quantity in physics is that it allows to switch from complicated surface and volume integrals to a point-wise description. The divergence is defined, for example, to write Maxwell's equations in a differential form. This is discussed in Chapter 3.

result is something finite:

$$\lim_{V \to 0} \frac{1}{V} \int_S \vec{F} \cdot d\vec{S} \equiv \nabla \cdot \vec{F}, \tag{A.59}$$

where S is the surface of the volume V and therefore also shrinks as we make the volume smaller. The resulting non-vanishing quantity $\nabla \cdot \vec{F}$ is known as the **divergence** of $\vec{F}$. In words, the divergence is the flux out of the volume V per unit volume in the limit that V becomes infinitesimally small. The definition in Eq. A.59 makes the interpretation of the divergence as flux density (Eq. A.56) precise. Take note that $\nabla \cdot \vec{F}$ is a scalar quantity, just like the flux ϕ.

We can understand the importance of this definition from a physical point of view. The divergence of a vector field is only non-zero if there is a source for the vector field present.[60] This is described by Gauss's law (Eq. 3.14):

[60] In electrodynamics, our sources are electric charges.

$$\oint_S \vec{E} \cdot d\vec{S} = \frac{1}{\epsilon_0} \int_V \rho dV = \frac{\rho_0}{\epsilon_0} V,$$

where we assumed, for simplicity, that the charge density is constant $\rho(\vec{x}) = \rho_0$. On the right-hand side we have the total charge contained in the volume V. Therefore, as we shrink the surface to zero, our volume shrinks to zero too. This means that

$$\lim_{S \to 0} \oint_S \vec{E} \cdot d\vec{S} = \lim_{V \to 0} \frac{\rho_0}{\epsilon_0} V = 0.$$

However, we can extract information about the vector field at a single point by getting rid of the volume on the right-hand side:[61]

[61] It's important to keep in mind that S is the surface of the volume V. Hence, if V shrinks to zero, its surface automatically shrinks to zero.

$$\lim_{V \to 0} \frac{1}{V} \oint_S \vec{E} \cdot d\vec{S} = \frac{\rho_0}{\epsilon_0}.$$

The construction appearing on the left-hand side is what we call the divergence and using the definition in Eq. A.59 our equation reads

$$\nabla \cdot E = \frac{\rho_0}{\epsilon_0}.$$

This is what we call Gauss' law in differential form (Eq. 3.16). For many practical purposes this differential form is a lot more useful than the integral form since it contains information about the field at single points and not only on surfaces.

Now, how is the strange looking definition of the divergence in Eq. A.59 related to the simple definition given at the beginning of this section?[62]

[62] Reminder:
$$\nabla \cdot \vec{F}(\vec{x}) = \partial_x F_1(\vec{x}) + \partial_y F_2(\vec{x}) + \partial_z F_3(\vec{x}).$$

To understand this, we consider the flux through a small rectangular parallelepiped explicitly and take the limit where the volume of the parallelepiped shrinks to zero.[63] We assume that the parallelepiped has edges of length $\Delta x, \Delta y, \Delta z$ parallel to the coordinate axes.

[63] Take note since we are really interested in the limit $\Delta V \to 0$, the actual shape of the volume doesn't matter and we simply choose a convenient one.

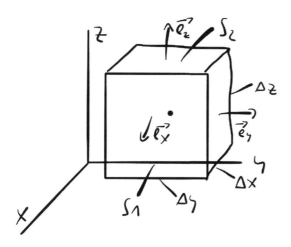

The flux through this parallelepiped is given by the sum over the flux through the six sides. We start with the face that we call S_1, as indicated in the figure above. The flux through S_1 is given by

$$\int_{S_1} \vec{F} \cdot \vec{n} dS = \int_{S_1} \vec{F} \cdot \vec{e}_x dS = \int_{S_1} F_x dS. \quad (A.60)$$

Since our parallelepiped is small, the integral is approximately equal to the value of $\vec{F}$ at the middle of the side times the area of the side:[64]

$$\int_{S_1} \vec{F} \cdot \vec{n} dS \simeq F_x(x + \frac{\Delta x}{2}, y, z) \Delta y \Delta z. \quad (A.61)$$

Analogously, the flux through the opposite side S_2 is[65]

$$\int_{S_2} \vec{F} \cdot \vec{n} dS \simeq -F_x(x - \frac{\Delta x}{2}, y, z) \Delta y \Delta z. \quad (A.62)$$

The total flux through the two sides S_1 and S_2 is therefore[66]

[64] The coordinates of the middle of the side S_1 are $(x + \frac{\Delta x}{2}, y, z)$ and the area of S_1 is $\Delta y \Delta z$. Moreover, take note that in the limit $V \to 0$ the formula we derive using this approximation becomes exact.

[65] The normal vector always points outwards and we therefore have $-\vec{e}_x$ as our normal vector. Moreover, the coordinates of the center of S_2 are $(x - \frac{\Delta x}{2}, y, z)$.

[66] We will talk about the contributions from the remaining four sides in a moment.

$$\int_{S_1+S_2} \vec{F} \cdot \vec{n} dS = \int_{S_1} \vec{F} \cdot \vec{n} dS + \int_{S_2} \vec{F} \cdot \vec{n} dS$$

⟩ Eq. A.61 and Eq. A.62

$$\simeq F_x(x+\frac{\Delta x}{2},y,z)\Delta y \Delta z - F_x(x-\frac{\Delta x}{2},y,z)\Delta y \Delta z$$

⟩ $\frac{\Delta x}{\Delta x} = 1$

$$= \frac{F_x(x+\frac{\Delta x}{2},y,z) - F_x(x-\frac{\Delta x}{2},y,z)}{\Delta x} \Delta x \Delta y \Delta z$$

⟩ $\Delta x \Delta y \Delta z = \Delta V$

$$= \frac{F_x(x+\frac{\Delta x}{2},y,z) - F_x(x-\frac{\Delta x}{2},y,z)}{\Delta x} \Delta V .$$

Once more we can see that the flux vanishes for $\Delta V \to 0$. However, the quantity

$$\frac{1}{\Delta V}\int_{S_1+S_2} \vec{F} \cdot \vec{n} dS \simeq \frac{F_x(x+\frac{\Delta x}{2},y,z) - F_x(x-\frac{\Delta x}{2},y,z)}{\Delta x}$$

is non-vanishing in the limit $\Delta V \to 0$:

$$\lim_{\Delta V \to 0}\frac{1}{\Delta V}\int_{S_1+S_2} \vec{F} \cdot \vec{n} dS = \lim_{\Delta V \to 0} \frac{F_x(x+\frac{\Delta x}{2},y,z) - F_x(x-\frac{\Delta x}{2},y,z)}{\Delta x}$$

⟩ if $\Delta V \to 0$, $\Delta x \to 0$

$$= \lim_{\Delta x \to 0} \frac{F_x(x+\frac{\Delta x}{2},y,z) - F_x(x-\frac{\Delta x}{2},y,z)}{\Delta x}$$

⟩

$$= \frac{\partial F_x}{\partial x} .$$

Here we used that the quantity appearing on the right-hand side is exactly the definition of the partial derivative in terms of the difference quotient. This is the flux per unit volume through the two sides S_1 and S_2 in the limit $\Delta V \to 0$.

As mentioned above, the total flux through the parallelepiped is given by the sum over the flux through all six sides. The calculations for the remaining four sides are completely analogous and yield $\frac{\partial F_y}{\partial y}$ plus $\frac{\partial F_z}{\partial z}$. The total flux per unit volume is therefore

$$\lim_{\Delta V \to 0} \frac{1}{\Delta V}\int_S \vec{F} \cdot \vec{n} dS = \frac{\partial F_x}{\partial x} + \frac{\partial F_y}{\partial y} + \frac{\partial F_z}{\partial z} . \qquad (A.63)$$

We can see, as promised above, that the two definitions of the divergence are really equivalent. In practice, of course, the simple formula in terms of derivatives is usually used.

Example: divergence

The divergence of the vector field

$$\vec{A}(\vec{x}) = \begin{pmatrix} x^2 y \\ yz \\ yz^2 \end{pmatrix} \tag{A.64}$$

is

$$\nabla \cdot \vec{A}(\vec{x}) = \begin{pmatrix} \partial_x \\ \partial_y \\ \partial_z \end{pmatrix} \cdot \begin{pmatrix} x^2 y \\ yz \\ yz^2 \end{pmatrix}$$

$$= \partial_x \left(x^2 y \right) + \partial_y \left(yz \right) + \partial_z \left(yz^2 \right)$$

$$= 2xy + z + 2yz. \tag{A.65}$$

A.12 Curl

[67] Reminder: × denotes the cross product.

The curl $\nabla \times$ is an operator which transforms a vector function $\vec{F}(\vec{x}) = (F_x(x), F_y(x), F_z(x))^T$ into a different vector function:[67]

$$\nabla \times \vec{F}(\vec{x}) = \begin{pmatrix} \partial_x \\ \partial_y \\ \partial_z \end{pmatrix} \times \begin{pmatrix} F_x(\vec{x}) \\ F_y(\vec{x}) \\ F_z(\vec{x}) \end{pmatrix} = \begin{pmatrix} \partial_y F_z(\vec{x}) - \partial_z F_y(\vec{x}) \\ \partial_z F_x(\vec{x}) - \partial_x F_z(\vec{x}) \\ \partial_x F_y(\vec{x}) - \partial_y F_x(\vec{x}) \end{pmatrix}$$

The meaning of the resulting vector function, called the curl, can be summarized as follows:[68]

[68] We discussed the circulation of a vector field in Appendix A.7. The idea of a circulation at a single point may seem strange at first glance. However, in principle, the idea is completely analogous to the idea behind the divergence of a vector field which describes the flux at a single point. Moreover, to understand it, imagine a little paddle wheel at each location. At locations with non-zero curl, the paddle wheel will rotate.

The curl of a vector function is a vector function $\vec{C}(\vec{x}) = \nabla \times \vec{F}(\vec{x})$ which describes the circulation $\vec{C}(\vec{r})$ of the vector field $\vec{F}(\vec{x})$ at a given point $\vec{r}$.

So, analogous to how the divergence $\nabla \cdot \vec{F}(\vec{x})$ tells us how flux is distributed through the system, the curl $\nabla \times \vec{F}(\vec{x})$ tells us how circulation is distributed throughout the system.

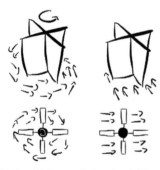

While the divergence is defined as flux per unit volume, the curl is defined as circulation per area:

$$\text{Curl} = \frac{\text{Circulation}}{\text{Area}}. \tag{A.66}$$

In this sense, an alternative name for the curl of a vector field could be "circulation density". [69]

[69] We will discuss below how this interpretation fits together with the definition in terms of derivatives given above.

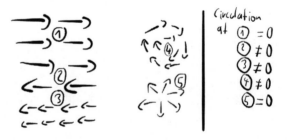

Let's talk about these definitions and interpretations in a bit more detail.

In Section A.7, we defined the circulation $\vec{C}(\vec{x})$ of a vector field $\vec{F}(\vec{x})$ along a path P by

$$\text{Circulation}(\vec{x}) = \oint_P \vec{F}(\vec{x}) \cdot \vec{t}\, ds, \qquad (A.67)$$

where s is a parameter which parameterizes our path and $\vec{t}(s)$ a unit vector tangential to the path.

Now, the idea is completely analogous to what we discussed in Appendix A.11. We shrink the path P until we are left with the neighborhood of a single point. However, if we simply take the limit $\lim_{P \to 0}$ in Eq. A.57, we find zero:

$$\lim_{P \to 0} \oint_P \vec{F} \cdot \vec{t}\, ds = 0. \qquad (A.68)$$

Therefore, this is not a useful quantity to describe the circulation at a single point. However, we can construct a useful quantity by dividing our circulation by the area S enclosed by the path and then take the limit[70]

$$\lim_{S \to 0} \frac{1}{S} \oint_P \vec{F} \cdot \vec{t}\, ds \neq 0. \qquad (A.69)$$

This quantity describes the circulation around the point enclosed by the specific path P.

[70] The line of thought here is completely analogous to what we discussed in Appendix A.11. Take note that since P is the path around S, the limit $\lim_{S \to 0}$ implies automatically that P shrinks to zero too.

However, there are many different paths around a given point and therefore Eq. A.69 doesn't contain all possible information about how the vector field $\vec{F}$ curls around the point in question. Instead, we need to calculate Eq. A.69 for multiple paths.

We must choose the paths in such a way that they can be used as basic building blocks for all possible paths. For example, in three dimensions we can use one path P_{yx} lying in the yx-plane (with normal vector $\vec{e}_z$), another one P_{xz} in the xz-plane (with normal vector $\vec{e}_y$) and a third one P_{yz} in the zy-plane (with normal vector $\vec{e}_x$). The circulation around these three paths contains complete information about how $\vec{F}$ curls.

In particular, it is useful to write the resulting three basic circulations as a vector

$$\vec{C}(\vec{x}) = \Big(\text{circulation around } P_{yz}\Big)\vec{e}_x + \Big(\text{circulation around } P_{xz}\Big)\vec{e}_y$$
$$+ \Big(\text{circulation around } P_{yx}\Big)\vec{e}_z$$
$$\equiv \nabla \times \vec{F}(\vec{x})$$

(A.70)

[71] For example, for $\vec{n} = \vec{e}_x$, we get the correct result for our path P_{yz}.

since we can then calculate the correct circulation around a general path with normal vector $\vec{n}$ using $\vec{n} \cdot \vec{C}(\vec{x})$.[71]

This idea allows us to derive the formula for the curl as a vector field involving derivatives, as stated at the beginning of this section.

Let's see how this works in practice.

We start with a small rectangular path P_{yx} in the yx-plane with edge lengths Δx and Δy.

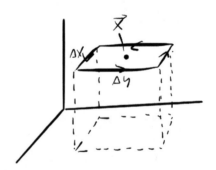

This path consist of four straight lines.

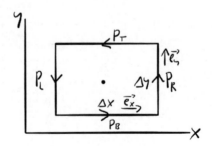

Therefore, to calculate the circulation along the path P_{yx}, we need to calculate the contributions from these four individual lines. We start with the contribution from P_B:

$$\int_{P_B} \vec{F} \cdot \vec{t}\, ds = \int_{P_B} F_x\, dx. \quad (A.71)$$

Since our rectangle is small, we can approximate the integral as the value of F_x at the center of the line times the length of the line[72]

$$\int_{P_B} \vec{F}(\vec{x}) \cdot \vec{e}_x\, ds = \int_{P_B} F_x(\vec{x})\, dx \simeq F_x\left(x, y - \frac{\Delta y}{2}, 0\right) \Delta x. \quad (A.72)$$

For the opposite line P_T, we can calculate analogously[73]

$$\int_{P_T} \vec{F}(\vec{x}) \cdot (-\vec{e}_x)\, ds = -\int_{P_T} F_x(\vec{x})\, dx \simeq -F_x\left(x, y + \frac{\Delta y}{2}, 0\right) \Delta x. \quad (A.73)$$

The total circulation along the two edges P_B and P_T is therefore[74]

$$\int_{P_B+P_T} \vec{F} \cdot \vec{t}\, ds = \int_{P_B} \vec{F} \cdot \vec{t}\, ds + \int_{P_T} \vec{F} \cdot \vec{t}\, ds$$

$$\simeq F_x\left(x, y - \frac{\Delta y}{2}, 0\right) \Delta x - F_x\left(x, y + \frac{\Delta y}{2}, 0\right) \Delta x$$

$$= \frac{F_x\left(x, y - \frac{\Delta y}{2}, 0\right) - F_x\left(x, y + \frac{\Delta y}{2}, 0\right)}{\Delta y} \Delta x \Delta y$$

$$= \frac{F_x\left(x, y - \frac{\Delta y}{2}, 0\right) - F_x\left(x, y + \frac{\Delta y}{2}, 0\right)}{\Delta y} \Delta S_{yx}. \quad (A.74)$$

[72] The coordinate of the center of the line P_B is $(x, y + \frac{\Delta y}{2}, 0)$ and the length of the line P_B is Δx. The vector tangential to the path is $\vec{e}_x$. Moreover, take note that in the limit $S \to 0$, the final formula will become exact.

[73] The coordinate of the center of the line P_T is $(x, y - \frac{\Delta y}{2}, 0)$ and the length of the line P_T is Δx. However, take note that the vector tangential to the path is now $-\vec{e}_x$.

[74] We will talk about the contributions from the remaining sides in a moment.

↷ Eq. A.72 and Eq. A.73

↷ $\frac{\Delta y}{\Delta y} = 1$

↷ $\Delta x \Delta y = \Delta S_{yx}$

We can see that the circulation vanishes for $\Delta S_{yx} \to 0$. However, the quantity

$$\frac{1}{\Delta S_{yx}} \int_{P_B+P_T} \vec{F} \cdot \vec{t}\, ds \simeq \frac{F_x\left(x, y - \frac{\Delta y}{2}, 0\right) - F_x\left(x, y + \frac{\Delta y}{2}, 0\right)}{\Delta y} \quad (A.75)$$

is non-vanishing in the limit $\Delta S_{yx} \to 0$:

$$\lim_{\Delta S_{yx} \to 0} \frac{1}{\Delta S_{yx}} \int_{P_B+P_T} \vec{F} \cdot \vec{t}\, ds = \lim_{\Delta S_{yx} \to 0} \frac{1}{\Delta S} \frac{F_x\left(x, y - \frac{\Delta y}{2}, 0\right) - F_x\left(x, y + \frac{\Delta y}{2}, 0\right)}{\Delta y}$$

$$= -\lim_{\Delta y \to 0} \frac{F_x\left(x, y + \frac{\Delta y}{2}, 0\right) - F_x\left(x, y - \frac{\Delta y}{2}, 0\right)}{\Delta y}$$

$$= -\frac{\partial F_x}{\partial y}.$$

Here we used that the quantity appearing on the right-hand side is exactly the definition of the partial derivative in terms of the difference quotient.

Completely analogously, we can calculate the contributions from the remaining two paths P_L and P_R:

$$\lim_{\Delta S_{yx} \to 0} \frac{1}{\Delta S_{yx}} \int_{P_L+P_R} \vec{F_y} \cdot \vec{t}\, ds = \frac{\partial F_y}{\partial x}. \qquad (A.76)$$

Therefore, the total circulation per unit area along the path P_{yx} is

$$\lim_{\Delta S_{yx} \to 0} \frac{1}{\Delta S_{yx}} \int_{P_{yx}} \vec{F} \cdot \vec{t}\, ds = \lim_{\Delta S \to 0} \frac{1}{\Delta S} \int_{P_B+P_T+P_L+P_R} \vec{F} \cdot \vec{t}\, ds$$

$$= \frac{\partial F_y}{\partial x} - \frac{\partial F_x}{\partial y}. \qquad (A.77)$$

Moreover, following exactly the same steps we can calculate the circulation per unit area along the paths P_{xz} and P_{yz}:

$$\lim_{\Delta S_{xz} \to 0} \frac{1}{\Delta S_{xz}} \int_{P_{xz}} \vec{F} \cdot \vec{t}\, ds = \frac{\partial F_x}{\partial z} - \frac{\partial F_z}{\partial x}$$

$$\lim_{\Delta S_{yz} \to 0} \frac{1}{\Delta S_{yz}} \int_{P_{yz}} \vec{F} \cdot \vec{t}\, ds = \frac{\partial F_z}{\partial y} - \frac{\partial F_y}{\partial z}. \qquad (A.78)$$

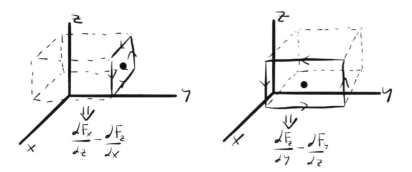

Therefore, using Eq. A.70, we can conclude that

$$\nabla \times \vec{F}(\vec{x}) = \lim_{\Delta S_{yz} \to 0} \lim_{\Delta S_{xz} \to 0} \lim_{\Delta S_{yx} \to 0} \begin{pmatrix} \frac{1}{\Delta S_{yz}} \int_{P_{yz}} \vec{F} \cdot \vec{t} \, ds \\ \frac{1}{\Delta S_{xz}} \int_{P_{xz}} \vec{F} \cdot \vec{t} \, ds \\ \frac{1}{\Delta S_{yx}} \int_{P_{yx}} \vec{F} \cdot \vec{t} \, ds \end{pmatrix} = \begin{pmatrix} \partial_y F_z(x) - \partial_z F_y(x) \\ \partial_z F_x(x) - \partial_x F_z(x)) \\ \partial_x F_y(x) - \partial_y F_x(x) \end{pmatrix}$$

which is exactly the formula given at the beginning of this chapter.

Example: curl

The curl of the vector field

$$\vec{A}(\vec{x}) = \begin{pmatrix} x^2 y \\ yz \\ yz^2 \end{pmatrix} \quad (A.79)$$

is

$$\nabla \times \vec{A}(\vec{x}) = \begin{pmatrix} \partial_x \\ \partial_y \\ \partial_z \end{pmatrix} \times \begin{pmatrix} x^2 y \\ yz \\ yz^2 \end{pmatrix}$$

$$= \begin{pmatrix} \partial_y (yz^2) - \partial_z (yz) \\ \partial_z (x^2 y) - \partial_x (yz^2) \\ \partial_x (yz) - \partial_y (x^2 y) \end{pmatrix}$$

$$= \begin{pmatrix} z^2 - y \\ 0 - 0 \\ 0 - x^2 \end{pmatrix}$$

$$= \begin{pmatrix} z^2 - y \\ 0 \\ -x^2 \end{pmatrix}. \quad (A.80)$$

A.13 The fundamental theorem for gradients

The fundamental theorem for gradients reads

$$\int_P \nabla \phi(\vec{x}) \cdot d\vec{s} = \phi(\vec{r}_1) - \phi(\vec{r}_0) \tag{A.81}$$

where $\vec{r}_0$ is the starting point and $\vec{r}_1$ the end point of the path P. Therefore, in words it tells us

> The path integral of the gradient of a scalar field is equal to the value of the field at the end point minus its value at the starting point.

In other words, it tells us that taking the gradient and calculating the path integral are inverse operations, analogous to how taking the derivative and integrating are inverse operations for ordinary functions.

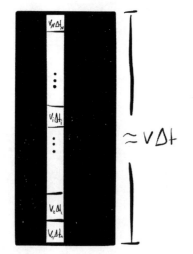

To understand the theorem, let's imagine we want to measure the height of a building. There are two possibilities:

▷ A rough approximation can be calculated by climbing the stairs and measuring the time that we need to get to the top Δt. We can then multiply this result by our average velocity to calculate the height of the building

$$\text{height} \simeq v\Delta t. \tag{A.82}$$

We get a better result, by measuring the time we need for each step Δx_i and then multiplying it by our velocity v_i during this step. This yields the height of each step and the total height is therefore

$$\text{height} \simeq \sum_i^N v_i \Delta t_i, \tag{A.83}$$

where N denotes the number of steps. In the limit of an infinitesimal step size, the sum becomes an integral and we get an exact formula for the height of the building:[75]

$$\text{height} = \int_0^T v(t)dt = \int_0^T \frac{dx(t)}{dt}dt. \quad (A.84)$$

This is exactly the method used on the left-hand side in Eq. A.81.[76]

▷ Alternatively, we can use an altimeter to measure the height above sea level at the top and bottom of the building and subtract these two results:

$$\text{height} = x(\text{top}) - x(\text{bottom}). \quad (A.85)$$

This is exactly the method used on the right-hand side in Eq. A.81.

[75] Take note that, formally, this is the same result that we get when we use a ruler to measure the rise at each step Δx_i and then add all these results up:

$$\sum_i^N \Delta x_i \to \int_0^H dx = \int_0^T \frac{dx(t)}{dt}dt,$$

where H denotes the height of the building and T is the total time that we need to get to the top.

[76] The only difference is that in Eq. A.81 our function $\phi(\vec{x})$ does depend on x, y, z and not only on x. Hence, to get the total height, we need to take the contributions coming from x, y, z into account.

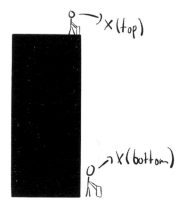

A.14 The fundamental theorem for divergences a.k.a. Gauss's theorem

The fundamental theorem for divergences reads

$$\int_V \nabla \cdot \vec{F}(\vec{x}) dV = \oint_S \vec{F}(\vec{x}) \cdot d\vec{S} \quad (A.86)$$

where S is the surface of the volume V. Therefore, in words it tells us

> The volume integral of the divergence of a vector field is equal to the integral of the field over the surface of the volume.

The theorem is important because it allows us to replace flux integrals with volume integrals and vice versa. This is useful, for example, if we want to rewrite Maxwell's equations in a more compact form.

We can understand in intuitive terms why the theorem is true as follows.

Let's imagine that the vector field $\vec{F}$ describes the flow of a particular quantity, say, an incompressible fluid like water. The flux integral on the right-hand side in Eq. A.86 gives us the total amount of water that flows through the surface S per unit time:[77]

$$\text{flux} = \oint_S \vec{F}(\vec{x}) \cdot d\vec{S}.$$

Now, analogous to how there are two ways of measuring the total height of a building, there is an alternative method to measure this total amount of water.[78]

To understand the second method, which is described by the left-hand side in Eq. A.86, we need to recall what the divergence

[77] We discussed the notion of "flux integral" in Appendix A.9.

[78] The two methods to measure the height of a building were discussed in Appendix A.13 in the context of the fundamental theorem for gradients.

of a vector field $\nabla \cdot \vec{F}(\vec{x})$ means. A non-zero divergence at a particular location always indicates that there is a source or a sink for the quantity in question.[79] If there is a water source inside the volume, a particular amount of water is necessarily forced out of the volume since the fluid is incompressible.

This means, that an alternative method to determine the amount of water flowing through the surface is to count the number of sources and sinks inside the volume.[80]

$$\text{flux} = \sum \text{sources (- sinks)} = \int_V \nabla \cdot \vec{F}(\vec{x}) \, dV$$

This is what the left-hand side in Eq. A.86 describes.

Alternatively, we can understand the left-hand side by recalling that the divergence of a vector field is defined as the flux density.[81] Hence, integrating the divergence over a volume yields the total flux.

[79] A non-zero divergence means that the vectors of the field spread out from the location in question. In physical terms, this means that there must be a source for the vector field. In the case of water, our source would be, for example, a faucet.

[80] A sink acts like a "negative source" and hence yields a negative contribution to the total flow. If there are more sinks than sources inside the volume, the total flow through the surface is negative, which means that water flows into the volume.

[81] This was discussed in Appendix A.11.

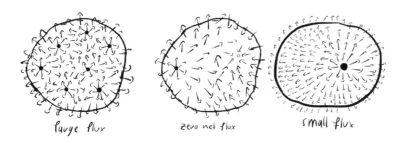

large flux   zero net flux   small flux

We can also understand this in more mathematical terms.

We start with the flux integral on the right-hand side in Eq. A.86

$$\oint_S \vec{F}(\vec{x}) \cdot d\vec{S}. \qquad (A.87)$$

The main idea is that we can divide the volume enclosed by the surface S into N subvolumes V_i.

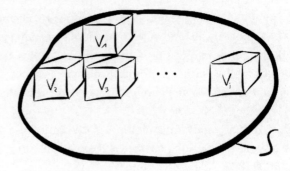

The total flux through the surface S is then equal to the flux through the surfaces of all subvolumes

$$\oint_S \vec{F}(\vec{x}) \cdot d\vec{S} \simeq \sum_i^N \int_{S_i} \vec{F}(\vec{x}) \cdot d\vec{S}. \qquad (A.88)$$

This is not obvious but we can understand this statement by considering the flux through the surfaces of two adjacent subvolumes

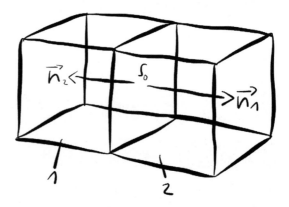

A crucial observation is that the flux through the common face S_0 of the two volumes cancels exactly. This happens because the flux is defined as the projection of the vector field onto the outward pointing normal vector.[82] The flux through the first subvolume therefore contains a contribution of the form

$$\int_{S_0} \vec{F}(\vec{x}) \cdot \vec{n}_1 dS, \qquad (A.89)$$

[82] This is discussed in Appendix A.9.

while the flux through the second subvolume contains a contribution of the form
$$\int_{S_0} \vec{F}(\vec{x}) \cdot \vec{n}_2 dS. \qquad (A.90)$$

Since $\vec{n}_2 = -\vec{n}_1$ it follows that [83]

$$\int_{S_0} \vec{F}(\vec{x}) \cdot \vec{n}_1 dS + \int_{S_0} \vec{F}(\vec{x}) \cdot \vec{n}_2 dS = \int_{S_0} \vec{F}(\vec{x}) \cdot \vec{n}_1 dS + \int_{S_0} \vec{F}(\vec{x}) \cdot (-\vec{n}_1) dS$$
$$\simeq 0. \qquad (A.91)$$

[83] In the limit of infinitesimal subvolumes, the two contributions cancel exactly.

We can therefore conclude that the flux through all common faces of the subvolumes cancel and the only non-vanishing contributions are those on the boundary of the volume. This is exactly the statement in Eq. A.88. However, this formula is only exact in the limit of infinitesimal subvolumes. To make sense of our formula in this limit, we modify it as follows[84]

[84] It will become clear in a moment why this modification is helpful.

$$\oint_S \vec{F}(\vec{x}) \cdot d\vec{S} \simeq \sum_i^N \int_{S_i} \vec{F}(\vec{x}) \cdot d\vec{S} \qquad \text{this is Eq. A.88}$$

$$\curvearrowright \frac{\Delta V_i}{\Delta V_i} = 1$$

$$= \sum_i^N \left(\frac{1}{\Delta V_i} \int_{S_i} \vec{F}(\vec{x}) \cdot d\vec{S} \right) \Delta V_i. \qquad (A.92)$$

If we now take the limit $\lim_{\Delta V_i \to 0}$ the expression between the large brackets is exactly the formal definition of the divergence (Eq. A.59). Moreover, the sum becomes an integral and we can therefore conclude that

$$\oint_S \vec{F}(\vec{x}) \cdot d\vec{S} = \int_V \nabla \cdot \vec{F}(\vec{x}) dV, \qquad (A.93)$$

which is exactly the fundamental theorem for divergences (Eq. A.86) stated at the beginning of this section.

A.15 The fundamental theorem for curls a.k.a. Stokes' theorem

The fundamental theorem for curls reads

$$\int_S \nabla \times \vec{F}(\vec{x}) \cdot d\vec{S} = \oint_P \vec{F}(\vec{x}) \cdot d\vec{l} \qquad (A.94)$$

where P is the boundary of the surface S. Therefore, in words it tells us

> The surface integral of the curl of a vector field is equal to the integral of the field over the boundary of the surface.

The theorem is important because it allows us to replace path integrals with surface integrals and vice versa. This is useful, for example, if we want to rewrite Maxwell's equations in a more compact form.

We can understand why the theorem is correct as follows.

The path integral on the right-hand side in Eq. A.94 describes the circulation of the vector field $\vec{F}(\vec{x})$.[85]

[85] The circulation of a vector field was discussed in Appendix A.7.

The curl on the left-hand side in Eq. A.94 is a measure for the circulation at a particular point. In other words, the curl describes a circulation density.[86] Hence, by integrating the curl over a particular area yields the total circulation.

[86] This was discussed in Appendix A.12.

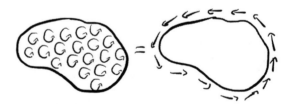

We can also understand this in more mathematical terms.

We start with the circulation integral on the right-hand side in Eq. A.94:[87]

[87] Take note that the steps we follow here are completely analogous to the steps we followed to derive the fundamental theorem for divergences in Appendix A.13.

$$\oint_P \vec{F}(\vec{x}) \cdot d\vec{l} \tag{A.95}$$

The main idea is that we can divide the surface enclosed by the path P into N small surfaces S_i.

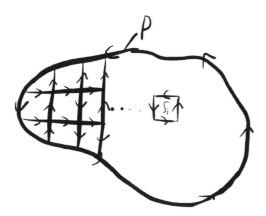

The total circulation along the path P is then equal to the circulation along the boundaries of all these small surfaces

$$\oint_P \vec{F}(\vec{x}) \cdot d\vec{l} \simeq \sum_i^N \oint_P \vec{F}(\vec{x}) \cdot d\vec{l}. \tag{A.96}$$

This is not obvious but we can understand this statement by considering the circulation along the boundary of two small surfaces.

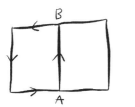 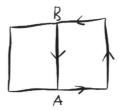

The crucial observation is that the circulation along the common edge cancels:

$$\int_A^B \vec{F}(\vec{x}) \cdot d\vec{l} + \int_B^A \vec{F}(\vec{x}) \cdot d\vec{l} = 0. \tag{A.97}$$

[88] We can understand that the actual shape of the surface doesn't matter. There are many possible surfaces with the same boundary path P. However, since the contributions from small rectangles on the surface cancel anyway, the exact shape doesn't matter.

[89] It will become clear in a moment why this modification is helpful.

We can therefore conclude that the circulation along all common edges vanishes and the only non-vanishing contributions are those on the boundary of the surface.[88] This is exactly the statement in Eq. A.96. This formula is only exact in the limit of infinitesimal surfaces $\lim_{\Delta S_i \to 0}$. Before we consider this limit, we rewrite our equation as follows[89]

$$\oint_P \vec{F}(\vec{x}) \cdot d\vec{l} \simeq \sum_i^N \oint_P \vec{F}(\vec{x}) \cdot d\vec{l} \qquad \text{this is Eq. A.96}$$

$$\frac{\Delta S_i}{\Delta S_i} = 1$$

$$= \sum_i^N \left(\frac{1}{\Delta S_i} \oint_P \vec{F}(\vec{x}) \cdot d\vec{l} \right) \Delta S_i . \qquad (A.98)$$

If we now take the limit $\lim_{\Delta S_i \to 0}$ the expression between the large brackets is exactly the formal definition of the curl (Eq. A.69). Moreover, the sum becomes an integral, and we can therefore conclude

$$\oint_P \vec{F}(\vec{x}) \cdot d\vec{l} = \int_S \nabla \times \vec{F}(\vec{x}) \cdot d\vec{S}, \qquad (A.99)$$

which is exactly the fundamental theorem for curls (Eq. A.94).

A.16 Vector identities

There are many useful mathematical relations involving the various quantities introduced in the previous sections. These so-called vector identities can often be used to make formulas shorter and calculations simpler.

For example, we have learned that we can construct a vector function $\vec{g}(\vec{x})$, called the gradient, from any given scalar function $\vec{g}(\vec{x}) = \nabla f(\vec{x})$. An important fact is that a vector function which is constructed this way is always curl-free:[90]

$$\nabla \times \vec{g} = \nabla \times \nabla f(\vec{x}) = 0. \qquad (A.100)$$

[90] We will see why this is true below. Take note that in contrast, a general vector field, i.e. one which isn't constructed as the gradient of a scalar field, can have a non-zero curl.

This observation is important in physics for the following reason.

The work W done by a particular force along a specific path P between two points A and B is given by

$$W = \int_P \vec{F} \cdot \vec{t} ds. \tag{A.101}$$

We can now check if $\nabla \times \vec{F}$ is zero. If yes, we know immediately that we can write $\vec{F}$ as the gradient of a scalar function:[91]

$$W = \int_P \vec{F} \cdot \vec{t} ds = \int_P \nabla f \cdot \vec{t} ds = f(B) - f(A) \tag{A.102}$$

where A and B denotes the starting and endpoint of the path P. This means that the work done between A and B does not depend on the particular path between these two points. In addition, if we consider a circular path we have $A = B$ and therefore find

$$W = \oint_P \vec{F} \cdot \vec{t} ds = \oint_P \nabla f \cdot \vec{t} ds = f(A) - f(A) = 0. \tag{A.103}$$

In words, this means that if the force can be written as the gradient of some potential, the work done along a circular path is always zero. Forces with this property are known as **conservative forces**, since no energy gets lost along the way.[92]

Now, Eq. A.100 tells us that we can write the force as the gradient of some potential if the curl vanishes. This means that for any given force $\vec{F}(\vec{x})$, we can check immediately whether it's conservative or not, simply by calculating the curl $\nabla \times \vec{F}(\vec{x})$. If the curl vanishes, we know that $\vec{F}(\vec{x})$ is a conservative force.[93]

To summarize:

$\boxed{\int_P \vec{F} \cdot \vec{t} ds \text{ is path independent}} \longleftrightarrow \boxed{\vec{F} = \nabla \phi} \longleftrightarrow \boxed{\nabla \times \vec{F} = 0}$

Now, why is Eq. A.100 true?

It's possible to check a vector identity like the one in Eq. A.100

[91] We use the fundamental theorem for gradients, which is completely analogous to the usual fundamental theorem $\int_a^b \partial_x f(x) \, dx = f(b) - f(a)$. In other words, we simply use that integration and differentiation "cancel" each other.

[92] If work is needed to move along a circular path, some energy has to go missing, e.g. in the form of friction.

[93] In physics, a vector function describing a conservative force $\vec{F}(\vec{x})$ can be thought of to be originating from a corresponding potential ϕ: $\vec{F} = -\nabla \phi$. Famous examples are the Lorentz force and the gravitational force.

simply by brute force:

$$\nabla \times \nabla f(\vec{x}) = \nabla \times \begin{pmatrix} \partial_x f(\vec{x}) \\ \partial_y f(\vec{x}) \\ \partial_z f(\vec{x}) \end{pmatrix}$$

$$= \begin{pmatrix} \partial_y \partial_z f(\vec{x}) - \partial_z \partial_y f(\vec{x}) \\ \partial_z \partial_x f(\vec{x}) - \partial_x \partial_z f(\vec{x}) \\ \partial_x \partial_y f(\vec{x}) - \partial_y \partial_x f(\vec{x}) \end{pmatrix}$$

↻ $\partial_x \partial_y = \partial_y \partial_x$ etc.

$$= \begin{pmatrix} \partial_z \partial_y f(\vec{x}) - \partial_z \partial_y f(\vec{x}) \\ \partial_z \partial_x f(\vec{x}) - \partial_z \partial_x f(\vec{x}) \\ \partial_x \partial_y f(\vec{x}) - \partial_x \partial_y f(\vec{x}) \end{pmatrix}$$

↻

$$= \begin{pmatrix} 0 \\ 0 \\ 0 \end{pmatrix} \qquad \square \qquad (A.104)$$

However, we can also understand it intuitively.

If the scalar field $f(\vec{x})$ describes the height of a given terrain, the gradient describes its slope. Equation A.100 then tells us that there is no circulation in the arrows describing the slope. This is certainly a sensible statement since we can't walk from A to B going uphill and then from B to A walk uphill again.[94] In other words, we can't go uphill both ways.

[94] Recall that the gradient vectors always point in the direction of the biggest slope. A situation with a circulating slope would represent a situation like those that are depicted in the famous paintings by M. C. Escher.

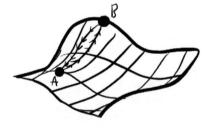

Alternatively, we can interpret Eq. A.100 for the case where the scalar field $f(\vec{x})$ describes the gravitational potential. The gradient $\nabla f(\vec{x})$ then describes the gravitational force. Now, Eq. A.100 states that a block slides downwards without spinning (at least in the absence of friction).

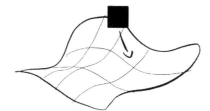

There is another particular important vector identity which we can use to understand the origin of the homogeneous Maxwell equations. This identity and its geometrical meaning are the topic of the next section.

A.16.1 The Bianchi identity

One of the homogeneous Maxwell equations reads $\nabla \cdot \vec{B} = 0$. In terms of the vector potential $\vec{A}$, we can write it as[95]

$$\nabla \cdot (\nabla \times \vec{A}) = 0. \tag{A.105}$$

[95] Reminder, Eq. 2.20:
$$\vec{B} = \nabla \times \vec{A}$$

An important observation is now that this equation is true for *any* vector field $\vec{A}$. In other words, the divergence of the curl is *always* zero for any vector field. This means that Eq. A.105 is a general vector identity analogous to $\nabla \times \nabla f(\vec{x}) = 0$ which we discussed in the previous section. Again, it's possible to check the validity of Eq. A.105 simply by brute force. However, there is also a different and more insightful way to see why Eq. A.105 is always true.

[96] We discussed Stokes' theorem in Appendix A.15 and Gauss's theorem in Appendix A.13.

The main idea is to integrate Eq. A.105 and then apply Stokes's theorem and Gauss's theorem[96]:

$$\int_V \nabla \cdot (\nabla \times \vec{A}) dV = \oint_{\delta V = S} (\nabla \times \vec{A}) \cdot d\vec{S} \quad \text{using Gauss's theorem Eq. A.86}$$
$$\quad \circlearrowright \text{ Stoke's theorem Eq. A.94}$$
$$= \oint_{\delta S} \vec{A} \cdot d\vec{l}$$
$$\quad \circlearrowright \quad \delta S = \delta \delta V = 0$$
$$= 0 \quad \checkmark \tag{A.106}$$

The crucial observation in the last step is that we integrate over the boundary of the surface of the volume. But the surface of a volume has no boundary and therefore the integral vanishes.

This is true in general. A boundary doesn't have a boundary. One way of understanding why this is true is by noting that every potential boundary point has already been used to define the boundary in the first place. For example, the boundary of a disk is a circle but a circle has no boundary.[97] Or, the boundary of a ball is a sphere but a sphere has no boundary.

To summarize: the deep reason why $\nabla \cdot (\nabla \times \vec{A}) = 0$ (Eq. A.105) is true is because the boundary of a boundary is always zero.[98].

[97] In contrast, a line has a boundary which consists of the two end points. However, this boundary (the two end points) do not have a boundary themselves.

[98] This interpretation of Bianchi identities is promoted mainly in [Misner, 1973]

There are many similar identities involving ∇, commonly known as **vector identities**. Below you'll find a list of particularly important vector identities. They can all be checked explicitly, analogous to what we did in Eq. A.104.

A.16.2 Summary of vector identities

$$\vec{\nabla} \cdot (\vec{\nabla} \times \vec{A}) \equiv \operatorname{div}(\operatorname{rot}\vec{A}) = (\vec{\nabla} \times \vec{\nabla}) \cdot \vec{A} \equiv 0$$

$$\vec{\nabla} \times (\vec{\nabla}\varphi) \equiv \operatorname{rot}\operatorname{grad}\varphi = (\vec{\nabla} \times \vec{\nabla})\varphi \equiv 0$$

$$\vec{\nabla} \cdot (\vec{A}\varphi) = \varphi \vec{\nabla} \cdot \vec{A} + \vec{A} \cdot \vec{\nabla}\varphi \quad \Longleftrightarrow \quad \operatorname{div}(\vec{A}\varphi) = \varphi \operatorname{div}\vec{A} + \vec{A} \cdot \operatorname{grad}\varphi$$

$$\vec{\nabla} \times (\vec{A}\varphi) = \varphi \vec{\nabla} \times \vec{A} - \vec{A} \times \vec{\nabla}\varphi \quad \Longleftrightarrow \quad \operatorname{rot}(\vec{A}\varphi) = \varphi \operatorname{rot}\vec{A} - \vec{A} \times \operatorname{grad}\varphi$$

$$\vec{\nabla} \cdot (\vec{A} \times \vec{B}) = \vec{B} \cdot (\vec{\nabla} \times \vec{A}) - \vec{A} \cdot (\vec{\nabla} \times \vec{B}) \quad \Longleftrightarrow \quad \operatorname{div}(\vec{A} \times \vec{B}) = \vec{B} \cdot \operatorname{rot}\vec{A} - \vec{A} \cdot \operatorname{rot}\vec{B}$$

$$\vec{\nabla} \times (\vec{A} \times \vec{B}) = (\vec{B} \cdot \vec{\nabla})\vec{A} - (\vec{A} \cdot \vec{\nabla})\vec{B} + \vec{A}(\vec{\nabla} \cdot \vec{B}) - \vec{B}(\vec{\nabla} \cdot \vec{A})$$

$$\Longleftrightarrow \quad \operatorname{rot}(\vec{A} \times \vec{B}) = (\vec{B}\operatorname{grad})\vec{A} - (\vec{A}\operatorname{grad})\vec{B} + \vec{A}(\operatorname{div}\vec{B}) - \vec{B}(\operatorname{div}\vec{A})$$

$$\vec{\nabla}(\vec{A} \cdot \vec{B}) = (\vec{B} \cdot \vec{\nabla})\vec{A} + (\vec{A} \cdot \vec{\nabla})\vec{B} + \vec{A} \times (\vec{\nabla} \times \vec{B}) + \vec{B} \times (\vec{\nabla} \times \vec{A})$$

$$\Longleftrightarrow \quad \operatorname{grad}(\vec{A} \cdot \vec{B}) = (\vec{B} \cdot \operatorname{grad})\vec{A} + (\vec{A} \cdot \operatorname{grad})\vec{B} + \vec{A} \times \operatorname{rot}\vec{B} + \vec{B} \times \operatorname{rot}\vec{A}$$

$$\vec{\nabla} \cdot (\vec{\nabla}\varphi) \equiv \operatorname{div}(\operatorname{grad}\varphi) \equiv \Delta\varphi = \frac{\partial^2\varphi}{\partial x^2} + \frac{\partial^2\varphi}{\partial y^2} + \frac{\partial^2\varphi}{\partial z^2}, \quad \Delta = \text{Laplace Operator}$$

$$\vec{\nabla} \times (\vec{\nabla} \times \vec{A}) \equiv \operatorname{rot}(\operatorname{rot}\vec{A}) = \vec{\nabla}(\vec{\nabla} \cdot \vec{A}) - (\vec{\nabla} \cdot \vec{\nabla})\vec{A} \equiv \operatorname{grad}\operatorname{div}\vec{A} - \Delta\vec{A}$$

A.17 Index notation and Maxwell's equations

In this section, we will neglect the constant c to unclutter the notation. In other words, we work in "natural units" where $c = 1$. This is often extremely convenient for fundamental considerations. However, if you want to get some number you can compare to an experiment (which usually use SI-units) you have to remember to insert the speed of light c in a few places.

In this appendix, we discuss how Maxwell's equations (Eq. 1.1) can be written in terms of the electric and magnetic fields (Eq. 1.4). Before we can do that, we need to recall some definitions and talk about various conventions.[99]

[99] Don't get demotivated if not every step in the following calculations is perfectly clear. It simply takes some time to get used to the index notation used in special relativity. Moreover, the details are not really important for the purpose of this book and it is a perfectly valid approach to simply take the equivalence of Eq. 1.1 and Eq. 1.4 for granted. It probably makes more sense to check the equivalence later once you are more familiar with the notation used in special relativity.

First of all, take note that the field-strength tensor has in total $4 \times 4 = 16$ components.[100] But not all these components are independent. This follows because the field-strength tensor $F_{\mu\nu}$ is antisymmetric

$$F_{\mu\nu} = -F_{\nu\mu},$$

which we can see by looking at the definition in terms of the potential

$$F_{\mu\nu} = \partial_\mu A_\nu - \partial_\nu A_\mu . \tag{A.107}$$

[100] Reminder: Greek indices like μ, ν or σ are always summed from 0 to 3: $x_\mu y_\mu = \sum_{\mu=0}^{3} x_\mu y_\mu$.

An antisymmetric (4×4) matrix has only 6 independent components and it is conventional to label the independent components as follows

$$F_{\mu\nu} = \begin{pmatrix} F_{00} & F_{01} & F_{02} & F_{03} \\ F_{10} & F_{11} & F_{12} & F_{13} \\ F_{20} & F_{21} & F_{22} & F_{23} \\ F_{30} & F_{31} & F_{32} & F_{33} \end{pmatrix}$$

$$\equiv \begin{pmatrix} 0 & -E_1 & -E_2 & -E_3 \\ E_1 & 0 & -B_3 & B_2 \\ E_2 & B_3 & 0 & -B_1 \\ E_3 & -B_2 & B_1 & 0 \end{pmatrix}.$$

[101] Reminder: Roman indices always run from 1 to 3.

In index notation, this means that[101]

$$F_{i0} = E_i . \tag{A.108}$$

For example, $F_{10} = E_1$ or $F_{20} = E_2$. The remaining three independent components, are in index notation, given by

$$F_{ij} = -\epsilon_{ijk} B_k. \tag{A.109}$$

For example,[102]

$$\begin{aligned}
F_{12} &= -\epsilon_{12k} B_k \\
&= -\underbrace{\epsilon_{121}}_{=0} B_1 - \underbrace{\epsilon_{122}}_{=0} B_2 - \underbrace{\epsilon_{123}}_{=1} B_3 \\
&= -B_3.
\end{aligned} \tag{A.110}$$

[102] Reminder: whenever an index appears twice in a term, an implicit sum is assumed. This is known as Einstein's summation convention.

Before we can rewrite Maxwell's equations in terms of the electric and magnetic fields, there is one more thing that we need to talk about.

Whenever an index appears twice but one time as a superscript and one time as a subscript, as it is the case in Maxwell's equations, there is an implicit minus sign between the 0-term and the remaining terms:[103]

$$\begin{aligned}
x_\mu y^\mu &= \sum_{\mu=0}^{3} x_\mu y_\mu \\
&= x_0 y_0 - \sum_{i=1}^{3} x_i y_i \\
&= x_0 y_0 - x_1 y_1 - x_2 y_2 - x_3 y_3.
\end{aligned}$$

[103] Don't worry if you don't understand the following explanations immediately. It takes some time getting used to this Minkowski notation. For us it is enough to know how to interpret a term of the form $x_\mu y^\mu$ like it appears in Maxwell's equations when we write them in terms of the field-strength tensor.

The reason for this convention is that the scalar product in special relativity reads[104]

$$x_\mu y_\mu \eta^{\mu\nu}, \tag{A.111}$$

where $\eta^{\mu\nu}$ is the Minkowski metric[105]

$$\eta^{\mu\nu} = \begin{pmatrix} 1 & 0 & 0 & 0 \\ 0 & -1 & 0 & 0 \\ 0 & 0 & -1 & 0 \\ 0 & 0 & 0 & -1 \end{pmatrix}. \tag{A.112}$$

[104] The scalar product is the correct method to combine two vectors to get something invariant, i.e. a scalar. The naive sum $x_0 y_0 + x_1 y_1 + x_2 y_2 + x_3 y_3$ is not invariant under Poincaré transformations and therefore not the correct scalar product. For some more information on special relativity, see Chapter 6.

[105] A metric is a mathematical tool which tells us the distance between two points. In the usual Euclidean space of classical mechanics, the metric is simply the unit matrix. However, in special relativity we are talking about distances in Minkowski spacetime and the correct way to measure distances is to use the Minkowski metric.

So we have

$$x_\mu \eta^{\mu\nu} y_\nu = \begin{pmatrix} x_0 & x_1 & x_2 & x_3 \end{pmatrix} \begin{pmatrix} 1 & 0 & 0 & 0 \\ 0 & -1 & 0 & 0 \\ 0 & 0 & -1 & 0 \\ 0 & 0 & 0 & -1 \end{pmatrix} \begin{pmatrix} y_0 \\ y_1 \\ y_2 \\ y_3 \end{pmatrix}$$

$$= x_0 y_0 - x_1 y_1 - x_2 y_2 - x_3 y_3. \tag{A.113}$$

To unclutter the notation, it is convenient to get rid of the Minkowski metric by introducing superscript indices:

$$y^\mu \equiv \eta^{\mu\nu} y_\nu. \tag{A.114}$$

We can then write the scalar product as

$$x_\mu y_\nu \eta^{\mu\nu} \equiv x_\mu y^\mu. \tag{A.115}$$

With this in mind, we can rewrite Maxwell's equations (Eq. 1.1)

$$\partial^\nu F_{\mu\nu} = \mu_0 J_\mu$$
$$\partial^\lambda F_{\mu\nu} + \partial^\mu F_{\nu\lambda} + \partial^\nu F_{\lambda\mu} = 0$$

in terms of the components E_i and B_i.

We start with the inhomogeneous Maxwell equations[106]

[106] The inhomogeneous Maxwell equations are those without a zero on the right-hand side.

$$\partial^\sigma F_{\rho\sigma} = J_\rho \tag{A.116}$$

For the three components $\rho \to i = 1, 2, 3$, we have

$$J_i = \partial^\sigma F_{i\sigma}$$
$$= \partial_0 F_{i0} - \partial_k F_{ik}$$
$$= \partial_0 E_i + \epsilon_{ikl} \partial_k B_l$$
$$\therefore \vec{J} = \partial_t \vec{E} + \nabla \times \vec{B}. \tag{A.117}$$

↪ Einstein's summation convention

↪ $F_{k0} = E_k$, Eq. A.108 and $F_{ik} = \epsilon_{ikl} B_l$, Eq. A.1⬤

↪ $\nabla \times \vec{B}$ is $\epsilon^{ikl} \partial_k B^l$ in index notation.

For the remaining component ($\rho \to 0$), we have

$$\begin{aligned}
J_0 &= \partial^\sigma F_{0\sigma} & &\circlearrowright \text{ Einstein's summation convention}\\
&= \partial_0 F_{00} - \partial_k F_{0k} & &\circlearrowright \; F_{00} = 0 \text{ and } F_{0k} = -F_{k0}\\
&= \partial_k F_{k0} & &\circlearrowright \; F_{k0} = E_k, \text{ Eq. A.108}\\
&= \partial_k E_k \, .
\end{aligned}$$
(A.118)

and we can conclude that

$$\nabla \cdot \vec{E} = J_0 \,. \qquad (A.119)$$

Analogously, we can rewrite the homogeneous Maxwell equations

$$\partial^\lambda F_{\mu\nu} + \partial^\mu F_{\nu\lambda} + \partial^\nu F_{\lambda\mu} = 0 \qquad (A.120)$$

in terms of E_i and B_i.

First of all, take note that we can write the homogeneous Maxwell equations as

$$\epsilon^{\mu\nu\lambda\delta} \partial^\lambda F_{\mu\nu} = 0 \,.$$

This follows when we multiply Eq. A.120 by $\epsilon^{\mu\nu\lambda\delta}$

$$\begin{aligned}
0 &= \epsilon^{\mu\nu\lambda\delta} \partial^\lambda F_{\mu\nu} + \epsilon^{\mu\nu\lambda\delta} \partial^\mu F_{\nu\lambda} + \epsilon^{\mu\nu\lambda\delta} \partial^\nu F_{\lambda\mu} & &\\
&= \epsilon^{\mu\nu\lambda\delta} \partial^\lambda F_{\mu\nu} + \epsilon^{\lambda\mu\nu\delta} \partial^\lambda F_{\mu\nu} + \epsilon^{\nu\lambda\mu\delta} \partial^\lambda F_{\mu\nu} & &\circlearrowright \text{ renaming indices}\\
&= \epsilon^{\mu\nu\lambda\delta} \partial^\lambda F_{\mu\nu} + \epsilon^{\mu\nu\lambda\delta} \partial^\lambda F_{\mu\nu} - \epsilon^{\mu\nu\lambda\delta} \partial^\lambda F_{\mu\nu} & &\circlearrowright \text{ switching indices of } \epsilon^{\lambda\mu\nu\delta}\\
&= \epsilon^{\mu\nu\lambda\delta} \partial^\lambda F_{\mu\nu} \;\checkmark & &\circlearrowright
\end{aligned}$$

We can then start by looking at the 0-component of this equation ($\delta \to 0$):[107]

[107] δ is the only free index, i.e. the only index which does not appear twice in each term.

$$0 = \epsilon^{\mu\nu\lambda 0} \partial^\lambda F_{\mu\nu}$$

⤷ $\epsilon^{\mu\nu\lambda\delta}$ is zero if two indices are equal

$$= \epsilon_{ijk} \partial_k F_{ij}$$

⤷ $F_{ij} = \epsilon_{ijl} B_l$, Eq. A.109

$$= \epsilon_{ijk} \partial_k (\epsilon_{ijl} B_l)$$

⤷

$$= \epsilon_{ijk} \epsilon_{ijl} \partial_k B_l$$

⤷ $\epsilon_{ijk}\epsilon_{ijl} = 2\delta_{kl}$, where δ_{kl} is the Kronecker delta

$$= 2\delta_{kl} \partial_k B_l$$

⤷

$$= 2\partial_l B_l \, .$$

We can therefore conclude that

$$\vec{\nabla} \cdot \vec{B} = 0 \, . \tag{A.121}$$

Analogously, we can take a look at the remaining components ($\delta \to i$) and derive

$$\vec{\nabla} \times \vec{E} + \partial_t \vec{B} = 0 \, . \tag{A.122}$$

This is the conventional form of the homogeneous Maxwell equation which is used for real-world applications.

A.17.1 Electrodynamical Lagrangian

In this appendix, we want to understand why we can write the electrodynamical Lagrangian in the following two forms:

$$L_{\text{Maxwell}} = \frac{1}{2}(\partial^\mu A^\nu \partial_\mu A_\nu - \partial^\mu A^\nu \partial_\nu A_\mu) = \frac{1}{4} F^{\mu\nu} F_{\mu\nu} \, . \tag{A.123}$$

We can see the equivalence as follows

$$L_{\text{Maxwell}} = \frac{1}{4} F^{\mu\nu} F_{\mu\nu}$$

$$= \frac{1}{4} (\partial^\mu A^\nu - \partial^\nu A^\mu)(\partial_\mu A_\nu - \partial_\nu A_\mu)$$

$$= \frac{1}{4} (\partial^\mu A^\nu \partial_\mu A_\nu - \partial^\mu A^\nu \partial_\nu A_\mu$$
$$\quad - \partial^\nu A^\mu \partial_\mu A_\nu + \partial^\nu A^\mu \partial_\nu A_\mu) \qquad \text{renaming dummy indices}$$

$$= \frac{1}{4} (\partial^\mu A^\nu \partial_\mu A_\nu - \partial^\mu A^\nu \partial_\nu A_\mu$$
$$\quad - \partial^\mu A^\nu \partial_\nu A_\mu + \partial^\mu A^\nu \partial_\mu A_\nu)$$

$$= \frac{1}{4} (2 \partial^\mu A^\nu \partial_\mu A_\nu - 2 \partial^\mu A^\nu \partial_\nu A_\mu)$$

$$= \frac{1}{2} (\partial^\mu A^\nu \partial_\mu A_\nu - \partial^\mu A^\nu \partial_\nu A_\mu) \quad \checkmark$$

$$\tag{A.124}$$

B

Taylor Expansion

The Taylor expansion is one of the most useful mathematical tools and we need it all the time in physics to simplify complicated systems and equations.

We can understand the basic idea as follows:

Imagine that you sit in your car and wonder what your exact location $l(t)$ will be in 10 minutes: $l(t_0 + 10 \text{ minutes}) = ?$

▷ A first guess is that your location will be exactly your *current* location
$$l(t_0 + 10 \text{ minutes}) \approx l(t_0).$$
Given how large the universe is and thus how many possible locations there are, this is certainly not too bad.

▷ If you want to do a bit better than that, you can also include your *current* velocity $\dot{l}(t_0) \equiv \partial_t l(t)\big|_{t_0}$.[1] The total distance you will travel in 10 minutes if you continue to move at your current velocity is this velocity times 10 minutes: $\dot{l}(t_0) \times 10 \text{ minutes}$. Therefore, your second estimate is your current location plus the velocity you are traveling times 10 minutes
$$l(t_0 + 10 \text{ minutes}) \approx l(t_0) + \dot{l}(t_0) \times 10 \text{ minutes}. \quad (B.1)$$

[1] Here ∂_t is a shorthand notation for $\frac{\partial}{\partial t}$ and $\partial_t l(t)$ yields the velocity (rate of change). After taking the derivative, we evaluate the velocity function $\dot{l}(t) \equiv \partial_t l(t)$ at t_0: $\dot{l}(t_0) = \partial_t l(t)\big|_{t_0}$.

▷ If you want to get an even better estimate you need to take into account that your velocity can possibly change. The rate of change of the velocity $\ddot{l}(t_0) = \partial_t^2 l(t)\big|_{t_0}$ is what we call acceleration. So in this third step, you additionally take your *current* acceleration into account[2]

$$l(t_0 + 10 \text{ minutes}) \approx l(t_0) + \dot{l}(t_0) \times 10 \text{ minutes}$$
$$+ \frac{1}{2}\ddot{l}(t_0) \times (10 \text{ minutes})^2.$$

▷ Our estimate will still not yield the perfect final location since, additionally, we need to take into account that our acceleration could change during the 10 minutes. We could therefore additionally take the current rate of change of our acceleration into account.

This game never ends and the only limiting factor is how precisely we want to estimate our future location. For many real-world purposes, our first order approximation (Eq. B.1) would already be perfectly sufficient.

The procedure described above is exactly the motivation behind the Taylor expansion. In general, we want to estimate the value of some function $f(x)$ at some value of x by using our knowledge of the function's value at some fixed point a. The **Taylor series** then reads[3]

$$f(x) = \sum_{n=0}^{\infty} \frac{f^{(n)}(a)(x-a)^n}{n!}$$
$$= \frac{f^{(0)}(a)(x-a)^0}{0!} + \frac{f^{(1)}(a)(x-a)^1}{1!} + \frac{f^{(2)}(a)(x-a)^2}{2!}$$
$$+ \frac{f^{(3)}(a)(x-a)^3}{3!} + \dots, \qquad (B.2)$$

where $f(a)$ is the value of the function at the point a we are expanding around. Moreover, $x - a$ is analogous to the 10 minute timespan we considered above. If we want to know the location at $x = 5{:}10$ pm by using our knowledge at $a = 5{:}00$ pm, we get $x - a = 5{:}10$ pm $- 5{:}00$ pm $= 10$ minutes. Therefore, this equation is completely analogous to our estimate of the future location we considered previously.

---

[2] The factor $\frac{1}{2}$ and that we need to square the 10 minutes follows since, to get from an acceleration to a location, we have to integrate twice:

$$\int dt \int dt \ddot{x}(t_0) =$$
$$\int dt \ddot{x}(t_0) t =$$
$$\frac{1}{2}\ddot{x}(t_0) t^2$$

where $\ddot{x}(t_0)$ is the value of the acceleration at $t = t_0$ (= a constant).

[3] Here the superscript n denotes the n-th derivative. For example $f^{(0)} = f$ and $f^{(1)}$ is $\partial_x f$.

To understand the Taylor expansion a bit better, it is helpful to look at concrete examples.

We start with one of the simplest but most important examples: the exponential function. Putting $f(x) = e^x$ into Eq. B.2 yields

$$e^x = \sum_{n=0}^{\infty} \frac{(e^0)^{(n)}(x-0)^n}{n!}$$

The crucial puzzle pieces that we need are therefore $(e^x)' = e^x$ and $e^0 = 1$. Putting this into the general formula (Eq. B.2) yields

$$e^x = \sum_{n=0}^{\infty} \frac{e^0(x-0)^n}{n!} = \sum_{n=0}^{\infty} \frac{x^n}{n!} \qquad (B.3)$$

This result can be used as a definition of e^x.

Next, let's assume that the function we want to approximate is $\sin(x)$ and we want to expand it around $x = 0$. Putting $f(x) = \sin(x)$ into Eq. B.2 yields

$$\sin(x) = \sum_{n=0}^{\infty} \frac{\sin^{(n)}(0)(x-0)^n}{n!}$$

The crucial information we therefore need is $(\sin(x))' = \cos(x)$, $(\cos(x))' = -\sin(x)$, $\cos(0) = 1$ and $\sin(0) = 0$. Because $\sin(0) = 0$, every term with even n vanishes, which we can use if we split the sum. Observe that

$$\sum_{n=0}^{\infty} n = \sum_{n=0}^{\infty} (2n+1) + \sum_{n=0}^{\infty} (2n)$$

$$1+2+3+4+5+6\ldots = 1+3+5+\ldots \quad +2+4+6+\ldots$$

$$(B.4)$$

Therefore, splitting the sum into even and odd terms yields

$$\sin(x) = \sum_{n=0}^{\infty} \frac{\sin^{(2n+1)}(0)(x-0)^{2n+1}}{(2n+1)!}$$
$$+ \sum_{n=0}^{\infty} \frac{\sin^{(2n)}(0)(x-0)^{2n}}{(2n)!}$$
$$= \sum_{n=0}^{\infty} \frac{\sin^{(2n+1)}(0)(x-0)^{2n+1}}{(2n+1)!}. \qquad (B.5)$$

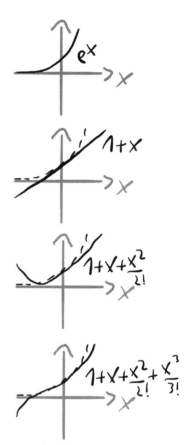

Moreover, every even derivative of $\sin(x)$ (i.e., $\sin^{(2n)}$) is again $\sin(x)$ or $-\sin(x)$. Therefore, the second term vanishes since $\sin(0) = 0$. The remaining terms are odd derivatives of $\sin(x)$, which are all proportional to $\cos(x)$. We now use

$$\sin(x)^{(1)} = \cos(x)$$
$$\sin(x)^{(2)} = \cos'(x) = -\sin(x)$$
$$\sin(x)^{(3)} = -\sin'(x) = -\cos(x)$$
$$\sin(x)^{(4)} = -\cos'(x) = \sin(x)$$
$$\sin(x)^{(5)} = \sin'(x) = \cos(x)$$

The general pattern is $\sin^{(2n+1)}(x) = (-1)^n \cos(x)$, as you can check by putting some integer values for n into the formula[4].

[4] $\sin^{(1)}(x) = \sin^{(2\cdot 0 + 1)}(x) = (-1)^0 \cos(x) = \cos(x)$, $\sin^{(3)}(x) = \sin^{(2\cdot 1 + 1)}(x) = (-1)^1 \cos(x) = -\cos(x)$

Thus, we can rewrite Eq. B.5 as

$$\sin(x) = \sum_{n=0}^{\infty} \frac{\sin^{(2n+1)}(0)(x-0)^{2n+1}}{(2n+1)!}$$

$$= \sum_{n=0}^{\infty} \frac{(-1)^n \cos(0)(x-0)^{2n+1}}{(2n+1)!} \quad \curvearrowright \cos(0) = 1$$

$$= \sum_{n=0}^{\infty} \frac{(-1)^n (x)^{2n+1}}{(2n+1)!}. \quad (B.6)$$

This is the Taylor expansion of $\sin(x)$, which we can also use as a definition of the sine function.

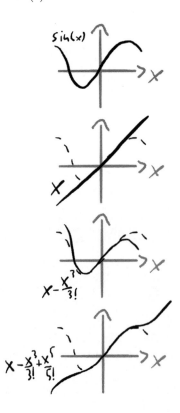

C

Delta Distribution

The delta distribution was invented as a tool that allows us to describe point sources. For example, in electrodynamics, an electron is a point source of the electromagnetic field. Usually, in electrodynamics, we describe the locations of charges using a quantity called charge density $\rho(\vec{x})$. A charge density encodes the amount of charge per unit volume. Hence, if we integrate it over some volume, we get the total charge contained in the volume[1]

$$\text{total charge inside } V = \int_V \rho(\vec{x}) dV. \tag{C.1}$$

[1] This is discussed in more detail in Section 2.2.

Now, how can we describe that there is only a single charge at one particular location? In other words: what's the charge density for a single point charge? We write the charge density of a point charge as

$$\rho_P(\vec{x}) = q\delta(\vec{x} - \vec{x}_0), \tag{C.2}$$

where q is the charge of the point charge, $\vec{x}_0$ its location and $\delta(\vec{x} - \vec{x}_0)$ the delta distribution. The defining property of the delta distribution is that *any* integral over a volume V_1 which contains the location of the point charge, yields exactly q:

$$\text{total charge inside } V_1 = \int_{V_1} \rho_P(\vec{x}) dV = \int_{V_1} q\delta(\vec{x} - \vec{x}_0) dV = q \tag{C.3}$$

but an integral over a different volume V_2 which does not contain the point $\vec{x}_0$ yields exactly zero:

$$\text{total charge inside } V_2 = \int_{V_2} \rho_p(\vec{x}) dV = \int_{V_2} q\delta(\vec{x} - \vec{x}_0) dV = 0. \tag{C.4}$$

This means that the $\delta(\vec{x} - \vec{x}_0)$ yields zero for all $\vec{x}$, except for $\vec{x} = \vec{x}_0$.

[2] The delta distribution is not really a function in the strict mathematical sense and therefore a new word was invented: distribution.

A good way to understand the delta distribution[2] (also known as the Dirac delta) is to recall a simpler but analogous mathematical object: the **Kronecker delta** δ_{ij}, which is defined as follows:

$$\delta_{ij} = \begin{cases} 1 & \text{if } i = j \\ 0 & \text{if } i \neq j \end{cases} \tag{C.5}$$

[3] For example, in two-dimensions

$$1_{(2\times 2)} = \begin{pmatrix} 1 & 0 \\ 0 & 1 \end{pmatrix}. \tag{C.6}$$

In matrix form, the Kronecker delta is simply the unit matrix[3]. The Kronecker delta δ_{ij} is useful because it allows us to pick one specific term of any sum. For example, let's consider the sum

$$\sum_{i=1}^{3} a_i b_j = a_1 b_j + a_2 b_j + a_3 b_j \tag{C.7}$$

and let's say we want to extract only the second term. We can do this by multiplying the sum by the Kronecker delta δ_{2i}:

$$\sum_{i=1}^{3} \delta_{2i} a_i b_j = \underbrace{\delta_{21}}_{=0} a_1 b_j + \underbrace{\delta_{22}}_{=1} a_2 b_j + \underbrace{\delta_{23}}_{=0} a_3 b_j = a_2 b_j. \tag{C.8}$$

In general, we have

$$\sum_{i=1}^{3} \delta_{ik} a_i b_j = a_k b_j. \tag{C.9}$$

The **delta distribution** $\delta(x - y)$ is a generalization of this idea for integrals instead of sums.[4] This means that we can use the delta distribution to extract specific terms from any given integral:[5]

$$\int dx f(x) \delta(x - y) = f(y). \quad \text{(C.10)}$$

[4] To unclutter the notation, we restrict ourselves to one-dimension.

[5] Take note that this implies the statement made above for $f(x) = q$:

$$\int dx q \delta(x - y) = q.$$

In words, this means that the delta distribution allows us to extract exactly one term - the term $x = y$ - from the infinitely many terms which we sum over as indicated by the integral sign. For example,

$$\int dx f(x) \delta(x - 2) = f(2).$$

Now, one example where the Kronecker delta appears is

$$\frac{\partial x_i}{\partial x_j} = \delta_{ij}. \quad \text{(C.11)}$$

The derivative of $\partial_x x = 1$, whereas $\partial_x y = 0$ and $\partial_x z = 0$.

Completely analogously, the delta distribution appears as follows:

$$\frac{\partial f(x_i)}{\partial f(x_j)} = \delta(x_i - x_j). \quad \text{(C.12)}$$

The delta distribution is also often introduced by the following definition:

$$\delta(x - y) = \begin{cases} \infty & \text{if } x = y, \\ 0 & \text{if } x \neq y \end{cases}, \quad \text{(C.13)}$$

which is somewhat analogous to the definition of the Kronecker delta in Eq. C.5. Moreover, when we use a constant function in Eq. C.10, for example, $f(x) = 1$, we get the following remarkable equation:

$$\int dx 1 \delta(x - y) = 1. \quad \text{(C.14)}$$

The thing is that Eq. C.10 tells us that if we have the delta distribution $\delta(x - y)$ together with a function under an integral, the result is the value of the function at $y = x$. Here, we have a constant function and its value at $y = x$ is simply 1.

In words, these properties mean that the delta distribution is an infinitely thin (only non-zero at $y = x$) and also an infinitely high function that yields exactly one if we integrate it all over space.

Bibliography

Y. Aharonov and D. Bohm. Significance of electromagnetic potentials in the quantum theory. *Phys. Rev.*, 115:485–491, 1959. DOI: 10.1103/PhysRev.115.485. [,95(1959)].

H. J. Bernstein and A. V. Phillips. Fiber Bundles and Quantum Theory. *Sci. Am.*, 245:94–109, 1981. DOI: 10.1038/scientificamerican0781-122.

Jennifer Coopersmith. *The lazy universe : an introduction to the principle of least action.* Oxford University Press, Oxford New York, NY, 2017. ISBN 9780198743040.

Richard Feynman. *The Feynman lectures on physics.* Addison-Wesley, San Francisco, Calif. Harlow, 2011. ISBN 9780805390650.

Daniel Fleisch. *A student's guide to Maxwell's equations.* Cambridge University Press, Cambridge, UK New York, 2008. ISBN 978-0521701471.

Daniel Fleisch. *A student's guide to vectors and tensors.* Cambridge University Press, Cambridge New York, 2012. ISBN 9781139031035.

A. P. French. *Special relativity.* Norton, New York, 1968. ISBN 9780393097931.

Juergen Freund. *Special relativity for beginners : a textbook for undergraduates.* World Scientific, Singapore, 2008. ISBN 9789812771599.

J. Gratus. A pictorial introduction to differential geometry, leading to Maxwell's equations as three pictures. *ArXiv e-prints*, September 2017.

David Griffiths. *Introduction to electrodynamics*. Pearson Education Limited, Harlow, 2014. ISBN 9781292021423.

Kirill Ilinski. Physics of Finance. 1997.

John Jackson. *Classical electrodynamics*. Wiley, New York, 1999. ISBN 9780471309321.

Juan Maldacena. The symmetry and simplicity of the laws of physics and the Higgs boson. *Eur. J. Phys.*, 37(1):015802, 2016. DOI: 10.1088/0143-0807/37/1/015802.

Charles Misner. *Gravitation*. W.H. Freeman and Company, New York, 1973. ISBN 9780716703440.

Edward Purcell. *Electricity and magnetism*. Cambridge University Press, Cambridge, 2013. ISBN 9781107014022.

H. M. Schey. *Div, grad, curl, and all that : an informal text on vector calculus*. W.W. Norton & Company, New York, 2005. ISBN 9780393925166.

Jakob Schwichtenberg. *Physics from Symmetry*. Springer, Cham, Switzerland, 2018a. ISBN 978-3319666303.

Jakob Schwichtenberg. *No-Nonsense Quantum Mechanics*. No-Nonsense Books, Karlsruhe, Germany, 2018b. ISBN 978-1719838719.

Jakob Schwichtenberg. *Physics from Finance*. No-Nonsense Books, Karlsruhe, Germany, 2019. ISBN 978-1795882415.

Gerard 't Hooft. Gauge theories of the forces between elementary particles. *Sci. Am.*, 242N6:90–116, 1980. [,78(1980)].

Edwin Taylor. *Spacetime physics : introduction to special relativity*. W.H. Freeman, New York, 1992. ISBN 9780716723271.

D. Wallace and Hilary Greaves. Empirical consequences of symmetries. *British Journal for the Philosophy of Science*, 65(1): 59–89, 2014.

K. Young. Foreign exchange market as a lattice gauge theory. *American Journal of Physics*, 67(10):862–868, 1999. DOI: 10.1119/1.19139. URL https://doi.org/10.1119/1.19139.

Andrew Zangwill. *Modern electrodynamics*. Cambridge University Press, Cambridge, 2013. ISBN 9780521896979.

Index

Ampere-Maxwell law, 79
amplitude, 145
angular frequency, 144
    spatial, 143
    temporal, 144

Bianchi identity, 171, 287

charge, 32
charge density, 35
    static, 95
circulation, 253
connection, 213
continuity equation, 54
Coulomb gauge, 178
Coulomb potential, 105
Coulomb's law, 104
cross product, 238
curl, 270
current, 37
curvature, 215

dipole, 110
dipole moment, 113
divergence, 264
dot product, 235

Einstein summation convention, 13
electric charge, 32
electric current, 37
    steady, 95

electric field, 44
    dipole, 110
    general charge distribution, 111
    point charge, 101
    sphere, 106
electric permittivity, 63
Electrodynamics, 133
electromagnetic field, 42
electromagnetic potential, 48
electromagnetic wave
    energy, 151
Electrostatics, 93, 95

Faraday's law, 74
fiber bundle, 212
field, 240
    scalar, 241
    vector, 241
field energy, 151
flux, 257
fundamental theorem
    calculus, 276
    curls, 281
    divergences, 278
    gradients, 276
fundamental theorem of calculus, 231

gauge connection, 213
    electrodynamics, 217
    quantum mechanics, 216
    toy model, 216

gauge group, 211
gauge symmetry, 175
   electrodynamics, 206
   quantum mechanics, 203
gauge theory, 181
   dynamics, 190
Gauss's law
   electric field, 61
   magnetic field, 69
gradient, 262
Greek indices, 13
Green's function, 121
group of gauge transformations, 211

Laplace equation, 121
line integral, 244
Lorentz force law, 21, 22, 57
   application, 115, 126

magnetic charge density, 69
magnetic field, 44
   wire, 123
magnetic monopole, 69
magnetic multipole moments, 131
Magnetostatics, 93, 95
magnets, 72
Maxwell's equations, 22
   covariant, 20
   homogeneous, 171
   index notation, 290
   inhomogeneous, 202
   origin, 169
multipole expansion, 114

Newton's second law, 115

path integral, 247
phase, 143
phase factor, 179
phase shift, 138
photons, 49
Poisson equation, 121
polarization, 145
   longitudinal, 146
   transversal, 146

redundancy, 188

scalar, 233
scalar field, 42, 241
special relativity, 163
spin, 72
summation convention, 13
superposition, 93, 111
surface integral, 254
symmetry, 182
   global, 183
   local, 183

tangent vector, 250
Taylor expansion, 112, 297
tensor, 233
   vector, 242
tensor field, 42, 242
test charge, 45
transformation
   active, 186
transformations
   passive, 186

vector, 233
vector calculus, 227
vector field, 42, 241
   solenoidal, 72
vector identities, 284

wave equation, 24, 82
   electric field, 83
   explicit solution, 134
   general solution, 138
   magnetic field, 83
wave equations
   monochromatic solutions, 139
   plane wave solutions, 139
   standing wave solutions, 140
wave function, 179
wave number, 143
wave vector, 143
wavelength, 144

Printed in Poland
by Amazon Fulfillment
Poland Sp. z o.o., Wrocław